T0140188

Studies in Big Data

Volume 61

Series Editor

Janusz Kacprzyk, Polish Academy of Sciences, Warsaw, Poland

The series "Studies in Big Data" (SBD) publishes new developments and advances in the various areas of Big Data- quickly and with a high quality. The intent is to cover the theory, research, development, and applications of Big Data, as embedded in the fields of engineering, computer science, physics, economics and life sciences. The books of the series refer to the analysis and understanding of large, complex, and/or distributed data sets generated from recent digital sources coming from sensors or other physical instruments as well as simulations, crowd sourcing, social networks or other internet transactions, such as emails or video click streams and other. The series contains monographs, lecture notes and edited volumes in Big Data spanning the areas of computational intelligence including neural networks, evolutionary computation, soft computing, fuzzy systems, as well as artificial intelligence, data mining, modern statistics and Operations research, as well as self-organizing systems. Of particular value to both the contributors and the readership are the short publication timeframe and the world-wide distribution, which enable both wide and rapid dissemination of research output.

** Indexing: The books of this series are submitted to ISI Web of Science, DBLP, Ulrichs, MathSciNet, Current Mathematical Publications, Mathematical Reviews, Zentralblatt Math: MetaPress and Springerlink.

More information about this series at http://www.springer.com/series/11970

Xiaolu Zhang · Kim-Kwang Raymond Choo
Editors

Digital Forensic Education

An Experiential Learning Approach

Springer

Editors
Xiaolu Zhang
Department of Information Systems
and Cyber Security
University of Texas at San Antonio
San Antonio, TX, USA

Kim-Kwang Raymond Choo
Department of Information Systems
and Cyber Security
University of Texas at San Antonio
San Antonio, TX, USA

ISSN 2197-6503 ISSN 2197-6511 (electronic)
Studies in Big Data
ISBN 978-3-030-23549-9 ISBN 978-3-030-23547-5 (eBook)
https://doi.org/10.1007/978-3-030-23547-5

© Springer Nature Switzerland AG 2020
This work is subject to copyright. All rights are reserved by the Publisher, whether the whole or part
of the material is concerned, specifically the rights of translation, reprinting, reuse of illustrations,
recitation, broadcasting, reproduction on microfilms or in any other physical way, and transmission
or information storage and retrieval, electronic adaptation, computer software, or by similar or dissimilar
methodology now known or hereafter developed.
The use of general descriptive names, registered names, trademarks, service marks, etc. in this
publication does not imply, even in the absence of a specific statement, that such names are exempt from
the relevant protective laws and regulations and therefore free for general use.
The publisher, the authors and the editors are safe to assume that the advice and information in this
book are believed to be true and accurate at the date of publication. Neither the publisher nor the
authors or the editors give a warranty, expressed or implied, with respect to the material contained
herein or for any errors or omissions that may have been made. The publisher remains neutral with regard
to jurisdictional claims in published maps and institutional affiliations.

This Springer imprint is published by the registered company Springer Nature Switzerland AG
The registered company address is: Gewerbestrasse 11, 6330 Cham, Switzerland

Foreword

Digital forensics is a science that has grown from the bottom-up, as opposed to the top-down approach that many sciences experience. It started because some police officer somewhere had a computer in evidence and needed to figure out how to extract evidentiary material off of it. As a result, many of the tools still used in digital forensics are home-grown, built by the practitioners that have to get at the evidence in a case. Although there are a lot of researchers working to develop new technologies and tools to extract and analyze evidence on digital devices, there are still a lot of unanswered questions in the field. As a result, practitioners have to be taught to think on their feet, and to solve problems thus far unseen in real cases. Students are taught to develop these problem-solving skills through hands-on exercises and practice. This experiential learning approach to digital forensics education is absolutely essential to the future understanding of digital systems and how they work.

In this book, Zhang and Choo take a look at the value of experiential learning through the use of student projects, including the digital forensic challenge offered by the Digital Forensic Research Workshop (DFRWS). Students were given a choice of trying to tackle the DFRWS challenge, or conduct research into some contemporary consumer technology. The DFRWS challenge is a competition offered every year leading up to the DFRWS in the late summer. It is always difficult and few people actually solve the challenge before the conference. This book describes the experiences gained from trying the DFRWS challenge and the gains made by the students who attempted it. It also provides feedback from the students on the knowledge gained during the semester, and the lessons learned from the hands-on experience encountered.

March 2019

David A. Dampier, Ph.D.
Professor and Department Chair
Information Systems and Cyber Security
College of Business
University of Texas at San Antonio
San Antonio, USA

Preface

Have you watched popular TV shows series such as CSI Cyber or NCIS? If yes, chances are you will have been introduced to digital forensics, or some parts of it. In its earlier days, digital forensics (or maybe it was known as computer forensics then) resulted from operational demands by the law enforcement community, for example, in the investigation of a crime that involves the use of some computing systems, typically a personal computer or server. In a typical forensic investigation, the forensic practitioner can preserve the scene by making a bit-by-bit copy of the storage medium, such as a hard disk in the personal computer, and/or other devices such as storage cards or copies of specific files if the system cannot be taken offline or is too big to copy. Forensic investigations are not limited to cyber criminal cases such as unauthorized access or attempts, accessing and disseminating of child abuse/exploitation materials. Increasingly, forensic investigations are also necessary in criminal activities involving the use of a computing/digital device (e.g., drug trafficking or money laundering cases where a mobile device is used for communication or fund transfer) and civil litigation (e.g., corporate espionage, and insider abuse of IT privileges).

A key component in the study of digital forensics is to establish best practices and models to ensure the reliability of the evidence, and practitioners need to examine all their methods and to expose them to external review to ensure the trust of the public. This highlights the importance of adequately training the next generation of digital forensic practitioners, which is the focus of this book. Specifically, in chapter "Experiential Learning in Digital Forensics" of this book, we describe how we expose students enrolled in the digital forensic courses (IS 4483 and IS 6363) to experiential learning opportunities, where they acquire the knowledge and skills of the subject-matter while also learning how to adapt to the ever-changing digital forensic landscape. We then present the reports detailing the findings of the student groups who worked on solving the Digital Forensic Research Workshop (DFRWS) forensic challenge in chapters "DFRWS IoT Forensic Challenge Report 1"–"DFRWS IoT Forensic Challenge Report 5" of this book. Student groups were also given the opportunity to conduct forensic research on a range of digital devices, as part of their semester-long project. Details of their findings on the forensic examination of Alexa/Google Home devices, Raspberry Pi (used in a man-in-the-middle attack), Kindle and Android devices, mobile devices, solid-state drives (SSDs), web browsers, microblogging sites, and social media, are presented in chapters "Alexa/Google Home Forensics"–"The Way Forward". Finally, we conclude this book with a number of potential research opportunities.

San Antonio, TX, USA

Xiaolu Zhang

March 2019

Kim-Kwang Raymond Choo

Acknowledgements

This book would not have been possible for the amazing students enrolled in our digital forensic classes, who were willing to dedicate their time and efforts on the final projects and share their contributions in this book (i.e., the authors of chapters "DFRWS IoT Forensic Challenge Report 1"–"Social Media Data in Digital Forensics Investigations"), as well as the student authors whose final projects were subsequently published either as peer-reviewed conference papers or journal papers.

We are also very fortunate to have extremely supportive colleagues, who go out of their way to support our efforts. Specifically, Dr. Nicole Lang Beebe generously shared her teaching materials and experience, as well as allowing us to build on her materials. Dr. David A. Dampier, the Department Chair, gave us the freedom to explore and implement experiential learning in our digital forensic courses. Mr. Rick White and his lab assistants were phenomenal in their support for our different requests (e.g., to install Mobile Phone Examiner Plus—MPE+ on several of the lab computers), often at extremely short notice.

We are also extremely grateful to Springer and their staff for their support in this project. It is not easy to keep on schedule but they were relentless ..., and clearly, in a positive way.

Contents

Experiential Learning in Digital Forensics

Xiaolu Zhang, Timothy T. Yuen and Kim-Kwang Raymond Choo

Abstract In this chapter, we introduce the concepts of digital forensics and experiential learning, and describe how we implement experiential learning in an undergraduate level digital forensic course and a graduate level digital forensic course. The students were given the option of working on either a digital forensic research topic or the Digital Forensic Research Workshop (DFRWS) IoT Forensic Challenge (2018–2019). We also report on the artifacts generated by the students working on both types of final projects.

1 Introduction

Digital forensics, one branch of forensic sciences,[1] play an increasingly important role in our society due to the prevalence of digital and computing devices. In digital forensics, the aim is generally to acquire courtroom evidence from digital devices (e.g., servers, personal computers, laptops, and mobile devices) that are used in some activity of interest, such as a cyber crime (e.g., unauthorized access, and intellectual property theft in corporate espionage) and physical crime (e.g., drug trafficking and money laundering).

Digital forensics have evolved since its earlier days of examining servers and personal computers. In recent times, the range and types of devices that can be the focus

---

[1] According to the American Academy of Forensic Sciences (AAFS; see https://www.aafs.org/about-aafs/sections/, Digital and Multimedia Sciences is been recognized as one of the 11 sub-disciplines in forensic sciences (Anthropology, Criminalistics, Engineering Sciences, General, Jurisprudence, Odontology, Pathology/Biology, Psychiatry and Behavioral Science, Questioned Documents, and Toxicology are the other ten forensic science sub-disciplines).

X. Zhang · T. T. Yuen · K.-K. R. Choo (✉)
University of Texas at San Antonio, San Antonio, TX 78249-0631, USA
e-mail: raymond.choo@fulbrightmail.org

X. Zhang
e-mail: xiaolu.zhang@utsa.edu

T. T. Yuen
e-mail: Timothy.Yuen@utsa.edu

© Springer Nature Switzerland AG 2020
X. Zhang and K.-K. R. Choo (eds.), *Digital Forensic Education*,
Studies in Big Data 61, https://doi.org/10.1007/978-3-030-23547-5_1

of a digital forensic investigation increase significantly as more of the 'things' around us become digitalized and/or Internet-connected (e.g., Internet of Everything), ranging from smart home devices (e.g., Amazon Echo, smart TVs and smart door bells with image acquisition features) to smart office/city (e.g., 3D printers, CCTVs and UGVs), and so on. As we become a more digital society, every person will create a much larger digital footprint tracked by more and more devices. For instance, an Amazon Echo device acted as a "witness" in an Arkansas Murder Case in 2016,[2] the data from a pacemaker was used in an arson case,[3] and the log from a Fitbit device was used to solve a murder case in 2017.[4]

Contemporary digital forensics can be broadly categorized into sub-branches or sub-disciplines, such as network forensics, memory forensics [9, 29], data and file carving [15, 25], and other device forensics (e.g., Internet of Things (IoT) [22, 23] and mobile devices [2, 14, 19, 30, 31]), cloud forensics [1, 10, 32], and anti-digital forensics (including reverse engineering) [5–7, 14, 27].

Digital forensic investigations are no longer restricted to criminal cases. Digital forensic investigations are also required in civil litigation [8, 17], which creates a societal need for digital forensic practitioners to serve a broader audience. However, as noted in the literature [18, 20, 21], it can be challenging for digital forensic practitioners to keep pace with technological advances, as well as the pressing need to train more digital forensic practitioners.

While the number of universities offering digital forensic programs are increasing, it remains small in comparison to related courses, such as those in cyber security and traditional forensic sciences. The University of Texas at San Antonio (UTSA), the authors' home institution, started offering digital forensic courses around 2008. We recognize the need to keep pace with technological and pedagogical advances, and are constantly refreshing our teaching materials to focus on contemporary and real-world practices. To prepare our students for industry, they are presented with experiential learning opportunities where they acquire the knowledge and skills of the subject-matter while also learning how to adapt to the ever-changing digital forensics landscape. Experiential learning engages students in active, hands-on, authentic learning activities through which they apply their knowledge and skills in real-world contexts [11, 12]. That is, what student learn is applied in meaningful activities that they would encounter in the actual field. Such experiential experiences have been shown to have positive academic outcomes in terms of higher order thinking in undergraduates: the more experiential experiences gained, the more valuable the students rated their own academic development [4]. Thus, students also find experiential learning useful.

This approach may involve instructors taking existing problems that students may encounter in industry and adapting them as class projects. In many cases, experiential

---

[2]https://www.forensicmag.com/news/2016/12/could-amazon-echo-be-witness-arkansas-murder-case.

[3]https://www.insurancefraud.org/IFNS-detail.htm?key=30365.

[4]https://www.forensicmag.com/news/2017/04/murdered-womans-fitbit-log-used-charge-husband.

Table 1 Courses and their research assignments

Course	Semester(s)	Research project	DFRWS challenge
IS 6363	Fall 2017, Fall 2018	Yes	No
IS 4483	Spring 2017, Spring 2018	Yes	No
IS 4483	Fall 2018	Yes	Yes

learning may take place in real-world environments, such as internships, service-learning, and other field-based contexts. The Kolb theory on experiential learning describes a model in which students transition between reflective observers to active experimenters while also transforming abstract concepts to concrete examples [12].

At UTSA, experiential learning was implemented in a graduate level digital forensic course (IS 6363: Computer Forensics) and an undergraduate level digital forensic course (IS 4483: Digital Forensic Analysis I) from Spring 2017 to Fall 2018. In order to facilitate hands-on learning and for the students to work on real-world problems, the authors (Choo and Zhang) introduced a semester-long group assignment worth 45% of the course marks. In IS 4483: Digital Forensic Analysis I offered in Fall 2018, the students were given the option of working on either a digital forensic research topic or the Digital Forensic Research Workshop (DFRWS) IoT Forensic Challenge (2018–2019).[5] In IS 6363: Computer Forensics and other offerings of IS 4483: Digital Forensic Analysis I, the students worked on a range of digital forensic research topics—see Table 1. A common aspect of experiential learning is the open-ended nature of its activities, which should foster and promote student creativity and engagement. Students have freedom in how they complete the activities, and there may be multiple solutions or no one "best" solution. Both the DFRWS and the research project options allowed students the choice and flexibility on what to investigate and how to investigate it. The active experimentation component of experiential learning allows students to test and re-test, and revise ideas they have. In that aspect, experiential learning is viewed as an ongoing, iterative learning process in which students are continually building upon prior knowledge [11].

For the digital forensic research project, students were provided several potential topics, as well as been encouraged to research, identify and discuss other potential topics with their instructors. Examples of topics provided in the class are as follows:

- **Cloud forensics**: What are the implications of cloud computing for forensic investigators, and/or challenges of cloud computing for digital forensics/electronic discovery (e.g. due to the lack of jurisprudence in applying the electronic discovery rules to third-party cloud computing data processing and retention services)? How and where can a forensic investigator acquire artifacts of forensic interest, say from a mobile device/application (app) that has been used to access the cloud service?
- **Smart mobile (anti-)forensics**: What are the forensic challenges associated with smart mobile devices (e.g. iPads, Android devices and BlackBerry devices)? What

---

[5]https://www.dfrws.org/dfrws-forensic-challenge.

types of artifacts can be forensically recovered from mobile devices, mobile apps, etc?

- **Big data**: What are the technical and non-technical challenges associated with big data?

Instructors serve as facilitators who provide guidance and some traditional direct instruction to students as they work through their projects [13]. The instructors set out tasks and assignments, grouped by milestones, to keep students on track. Both assignment types have four milestones, graded throughout the semester.

- **Milestone 1 (3%)**: A brief literature review methodology that describes how the literature will be surveyed (i.e., what literature review protocol will be used), as well as identification of the research question(s) that is/are supported by the introduction and literature review.
- **Milestone II (7%)**: Evidence creation (if applicable), preliminary evidence analysis and reporting of findings.
- **Milestone III (30%)**: In-depth analysis and reporting of findings, in the form of an academic paper (i.e., Introduction, Related work, Research methodology/setup, Findings and discussion, and Conclusion and future work).
- **Milestone IV (5%)**: Oral presentation to the class, followed by a question and answer session.

Throughout this book,[6] we present the artifacts generated by the students working on both types of final projects and intended to discuss the failure and success we gained during the past two years as we believe the achievement of the experiential learning in this book can be used for facilitating the practice-oriented digital forensic education in the future.

2   Project Design

In this section, we will explain how the experiential learning projects are designed.

2.1   DFRWS Digital Forensic Challenge

DFRWS Digital Forensic Challenge is an event organized by one of the key digital forensic conferences. Since the conference's inception in 2001, a number of digital forensic challenges have been conducted, such as memory forensics (2005), PlayStation 3 Forensics (2009), mobile malware forensics (2014), and SDN forensics (2016). Therefore, the most recent version of the forensic challenge, at the time of the course design and delivery, was chosen.

---

[6]In addition, several of the student groups' final projects were extended and published either as peer-reviewed conference papers [3, 16, 26] or peer-reviewed journal papers [24, 28].

Table 2 IoT devices in the challenge: a summary

No.	Device	Acquired data/image
1	Samsung galaxy edge S6	Physical extraction
2	iSmartAlarm	Diagnostic logs
		Memory image
3	QBee camera	N/A
4	Nest camera	N/A
5	Netgear Arlo base station	Memory image
		NVRAM settings
		TAR archive of the folder/tmp/media/nand
6	Netgear Arlo pro camera	N/A
7	Nest protect	N/A
8	Amazon echo	Extraction of cloud data obtained via CIFT
9	WinkHub	Filesystem TAR archive
10	Motion sensor	N/A
11	Door/Contact sensor	N/A

In addition to the data listed above, a Packet CAPture (PCAP) file of the local IoT network was provided

The scenario of this specific challenge is about the raid of an illegal drug lab, in which a SOHO IoT network was found. Table 2 summarizes the 11 IoT devices included in the challenge, as well as the types of data extracted from these IoT devices. Thus, with the given data, the students need to analyze these data and answer the following questions (from the Attorney General).

1. At what time was the illegal drug lab raided?
2. Could any of those involved in the raid by friends of the lab's owner?
3. How was the QBee camera disabled?

This assignment was then broken down into three milestones, each containing several tasks, to be achieved throughout the semester. Teams were provided with a digital forensic report template. For each milestone achieved, the groups need to complete the respective section(s) of the report. The groups were also encouraged to modify the template, as necessary.

Students are encouraged to adapt the structure suggested in Table 3 and modify it accordingly.

They were also briefed on the importance of crime scene reconstruction, as in conventional forensic science, crime scene reconstruction is a key step to gaining explicit knowledge of the case. In the digital forensic context, the network topology of the digital devices, for example, can be considered as the digital crime scene. Therefore, in Sect. 2 of the report, the students should attempt to reconstruct the network topology of the devices as part of their Milestone I, and to summarize the

Table 3 Structure of the report template

Section #	Suggested section heading	Related milestone
1	Case summary	Milestone III
1.1	Case breakdown	(None: Provided by the instructor)
1.2	Attorney's questions	(None: Provided by the instructor)
1.3	Brief answers to the questions	Milestone III
2	Case reconstruction	Milestones I to III
2.1	Digital crime scene reconstruction	Milestone I
2.2	Events reconstruction	Milestones II and III
3	Data analysis and major findings[a]	Milestones I to III
3.1	QBee camera	Milestones I to III
3.2	Nest camera	Milestones I to III
3.3	Arlo camera	Milestones I to III
3.4	iSmartAlarm	Milestones I to III
3.5	Nest protect	Milestones I to III
3.6	Amazon echo	Milestones I to III
3.7	Winkhub	Milestones I to III
4	Additional findings and assumptions (Optional)	Milestone III

[a]Background research of the given device should be undertaken as part of Milestone I, and digital forensic analysis should be completed as part of Milestone III

series of events (within a time-line) that are meaningful for the investigation as part of their Milestone III. Note that to have the information required to complete Sect. 2.1, the students need to analyze the provided PCAP file.

It is also suggested to have subsections within Sect. 3 to describe the IoT devices found in the crime scene, and the findings from the analysis of each device. Specifically, the groups need to report on the key evidences they found, the location of the evidences, the approaches/tools used (especially when the evidence was not stored in clear-text), the significance of the evidence, etc. Section 3 is a 'living document', in the sense that it will be updated constantly during the entire semester. However, the groups must include a brief introduction for each IoT device in the respective subsection as part of Milestone I. This is because we feel that it is important for the digital forensic examiners (students, in our context) to understand what the IoT devices are and how they behave in the IoT network (e.g. Arlo camera does not have an IP address, and it must connect to a Arlo base station).

To assist students with the three milestones, related lectures were delivered prior to the milestone due date and with sufficient time for the students to understand and apply the concepts (e.g., network forensics for milestone I, and mobile forensics for milestone II).

2.2 Research Projects

For students working on the research assignment, they would need to generate their own datasets for analysis. For example, students would need to decide on the (popular) app category (e.g., cloud storage, dating, or social networking) they wish to focus on, and then the actual app(s) they would be examining. Groups can decide to either focus on one specific app, but installed on different mobile devices with different operating systems (e.g., iPhone 6, iPhone 7, and different Android devices), or multiple devices running on mobile devices of a similar make and model. Browser forensics is another topic that students could work on, for example by examining the types of artefacts that could be recovered from the use of one or multiple browsers, in both normal browsing and private browsing modes, on different operating systems.

Once the groups had generated their datasets (typically over three or more weeks), then they started to perform their analysis (similar to the groups working on the forensic challenge).

The topics covered in both undergraduate and graduate courses are as follows:

- Overview of Digital Forensic Process
- Computer Foundations
- Volume/Partition Analysis
- File System Analysis
- FAT
- Mobile Forensics (Android and iOS Forensics)
- NTFS
- Deleted File Recovery and Recovery
- Signature Analysis/Data Carving
- Hashing, Hash Analysis
- Web Browsing Analysis
- Email Analysis
- String Search
- Cyber Law
- Forensic Preparation and Readiness, and Forensic Response.

3 Organization of the Book

The remaining of this book is structured as follows:

- Part I comprises five chapters, each describing the findings of a group's analysis of the forensic challenge.
- Part II comprises nine chapters, each describing the findings of a research topic chosen by the group.
- Chapter 16 concludes this book, and outlines a number of future research agenda.

References

1. Ahsan, M.M., Wahab, A.W.A., Idris, M.Y.I., Khan, S., Bachura, E., Choo, K.-K.R.: Class: cloud log assuring soundness and secrecy scheme for cloud forensics. IEEE Trans. Sustain. Comput. (2018)
2. Barmpatsalou, K., Cruz, T., Monteiro, E., Simões, P.: Current and future trends in mobile device forensics: a survey. ACM Comput. Surv. 51(3), 46:1–46:31 (2018). https://doi.org/10.1145/3177847
3. Cloyd, T., Osborn, T., Ellingboe, B., Glisson, W.B., Choo, K.R.: Browser analysis of residual facebook data. In: 17th IEEE International Conference on Trust, Security and Privacy in Computing and Communications/12th IEEE International Conference on Big Data Science and Engineering, Trustcom/Bigdatase 2018, New York, NY, USA, 1–3 August 2018, pp. 1440–1445 (2018). https://doi.org/10.1109/Trust-Com/BigDataSE.2018.00200
4. Coker, J.S., Heiser, E., Taylor, L., Book, C.: Impacts of experiential learning depth and breadth on student outcomes. J. Exp. Educ. 40(1), 5–23 (2017)
5. DOrazio, C., Ariffin, A., Choo, K.-K.R.: IoS anti-forensics: how can we securely conceal, delete and insert data? In: 2014 47th Hawaii International Conference on System Sciences, pp. 4838–4847 (2014)
6. D'Orazio, C.J., Choo, K.R.: Circumventing IoS security mechanisms for APT forensic investigations: a security taxonomy for cloud apps. Future Gener. Comput. Syst. 79, 247–261 (2018). https://doi.org/10.1016/j.future.2016.11.010
7. Eterovic-Soric, B., Choo, K.-K.R., Mubarak, S., Ashman, H.: Windows 7 antiforensics: a review and a novel approach. J. Forensic Sci. 62(4), 1054–1070 (2017)
8. Guo, H., Hou, J.: Review of the accreditation of digital forensics in China. Forensic Sci. Res. 3(3), 194–201 (2018)
9. Horst, L.V.D., Choo, K.R., Le-Khac, N.: Process memory investigation of the Bitcoin clients electrum and Bitcoin core. IEEE Access 5, 22385–22398 (2017). https://doi.org/10.1109/ACCESS.2017.2759766
10. Khan, S., Gani, A., Wahab, A.W.A., Bagiwa, M.A., Shiraz, M., Khan, S.U., Zomaya, A.Y.: Cloud log forensics: foundations, state of the art, and future directions. ACM Comput. Surv. 49(1), 7:1–7:42 (2016). https://doi.org/10.1145/2906149
11. Kolb, A.Y., Kolb, D.A.: Learning styles and learning spaces: enhancing experiential learning in higher education. Acad. Manag. Learn. Educ. 4(2), 193–212 (2005)
12. Kolb, A.Y., Kolb, D.A.: Experiential learning theory. In: Encyclopedia of the Sciences of Learning, pp. 1215–1219. Springer, Berlin (2012)
13. Konak, A., Clark, T.K., Nasereddin, M.: Using Kolb's experiential learning cycle to improve student learning in virtual computer laboratories. Comput. Educ. 72, 11–22 (2014)
14. Leom, M.D., Choo, K.-K.R., Hunt, R.: Remote wiping and secure deletion on mobile devices: a review. J. Forensic Sci. 61(6), 1473–1492 (2016)
15. Lin, X.: File carving. In: Introductory Computer Forensics, pp. 211–233. Springer, Berlin (2018)
16. Mata, N., Beebe, N., Choo, K.R.: Are your neighbors Swingers or Kinksters? Feeld app forensic analysis. In: 17th IEEE International Conference on Trust, Security and Privacy in Computing and Communications/12th IEEE International Conference on Big Data Science and Engineering, Trustcom/Bigdatase 2018, New York, NY, USA, 1–3 August 2018, pp. 1433–1439 (2018). https://doi.org/10.1109/Trust-Com/BigDataSE.2018.00199
17. My, A., Rights, C., Justice, S.: American bar association. Perspectives (2018)
18. Quick, D., Choo, K.-K.R.: Impacts of increasing volume of digital forensic data: a survey and future research challenges. Digit. Investig. 11(4), 273–294 (2014)
19. Quick, D., Choo, K.R.: Pervasive social networking forensics: intelligence and evidence from mobile device extracts. J. Netw. Comput. Appl. 86, 24–33 (2017). https://doi.org/10.1016/j.jnca.2016.11.018
20. Quick, D., Choo, K.R.: Big Digital Forensic Data - Volume 1: Data Reduction Framework and Selective Imaging. Springer, Berlin (2018). https://doi.org/10.1007/978-981-10-7763-0

21. Quick, D., Choo, K.R.: Big Digital Forensic Data - Volume 2: Quick Analysis for Evidence and Intelligence. Springer, Berlin (2018). https://doi.org/10.1007/978-981-13-0263-3
22. Quick, D., Choo, K.R.: IoT device forensics and data reduction. IEEE Access **6**, 47566–47574 (2018). https://doi.org/10.1109/ACCESS.2018.2867466
23. Rondeau, C.M., Temple, M.A., Lopez, J.: Industrial IoT cross-layer forensic investigation. In: Wiley Interdisciplinary Reviews: Forensic Science, p. e1322 (2019)
24. Shetty, R., Grispos, G., Choo, K.R.: Are you dating danger? an interdisciplinary approach to evaluating the (in)security of android dating apps. IEEE Trans. Sustain. Comput. (2019). https://doi.org/10.1109/TSUSC.2017.2783858
25. Shi, K., Xu, M., Jin, H., Qiao, T., Yang, X., Zheng, N., Choo, K.R.: A novel file carving algorithm for national marine electronics association (NMEA) logs in GPS forensics. Digit. Investig. **23**, 11–21 (2017). https://doi.org/10.1016/j.diin.2017.08.004
26. Smith, C., Dietrich, G., Choo, K.R.: Identification of forensic artifacts in VMWare virtualized computing. In: Security and Privacy in Communication Networks - Securecomm 2017 International Workshops, ATCS and Sepriot, Niagara Falls, Canada, 22–25 October 2017, Proceedings, pp. 85–103 (2017). https://doi.org/10.1007/978-3-319-78816-6_7
27. Stamm, M.C., Liu, K.J.R.: Anti-forensics of digital image compression. IEEE Trans. Inf. Forensics Secur. **6**(3–2), 1050–1065 (2011). https://doi.org/10.1109/TIFS.2011.2119314
28. Volety, T., Saini, S., McGhin, T., Liu, C.Z., Choo, K.R.: Cracking Bitcoin wallets: i want what you have in the wallets. Future Gener. Comput. Syst. **91**, 136–143 (2019). https://doi.org/10.1016/j.future.2018.08.029
29. Zhang, X., Hu, L., Song, S., Xie, Z., Meng, X., Zhao, K.: Windows volatile memory forensics based on correlation analysis. J. Netw. **9**(3), 645–653 (2014)
30. Zhang, X., Breitinger, F., Baggili, I.: Rapid android parser for investigating DEX files (RAPID). Digit. Investig. **17**, 28–39 (2016)
31. Zhang, X., Baggili, I., Breitinger, F.: Breaking into the vault: privacy, security and forensic analysis of android vault applications. Comput. Secur. **70**, 516–531 (2017)
32. Zhang, X., Grannis, J., Baggili, I., Beebe, N.L.: Frameup: an incriminatory attack on Storj: a peer to peer blockchain enabled distributed storage system. Digit. Investig. (2019)

Part I
Digital Forensic Challenge Case Studeis

DFRWS IoT Forensic Challenge Report 1

Daniel Palmer, Edward Blackburne and Theo Lemoine

Abstract In this chapter, we report on the findings of our group's analysis of the Digital Forensic Research Workshop (DFRWS) IoT Forensic Challenge (2018–2019).

1 Case Summary

1.1 Case Breakdown[1]

On 17 May 2018 at 10:40, the police were alerted that an illegal drug lab was invaded and unsuccessfully set on fire. The police respond promptly, and a forensic team is on scene at 10:45, including a digital forensic specialist.

The owner the illegal drug lab, Jessie Pinkman, is nowhere to be found. Police interrogate two of Jessie Pinkman's known associates: D. Pandana and S. Varga. Pandana and Varga admit having access to the drug lab's WiFi network but deny any involvement in the raid. They also say that Jessie Pinkman's had the IoT security systems installed because he feared attacks from a rival gang and that Jessie kept the alarm engaged in "Home" mode whenever he was inside the drug lab.

Within the drug lab the digital forensic specialist observes some IoT devices, including an alarm system (iSmartAlarm), three cameras (QBee Camera, Nest Camera and Arlo Pro) as well as a smoke detector (Nest Protect). An Amazon Echo and a WinkHub are also present.

---

[1] https://www.dfrws.org/dfrws-forensic-challenge.

D. Palmer (✉) · E. Blackburne · T. Lemoine
Department of Information System and Cyber Security,
University of Texas at San Antonio, San Antonio, TX, USA
e-mail: Daniel.SPalmer@yahoo.com

© Springer Nature Switzerland AG 2020
X. Zhang and K.-K. R. Choo (eds.), *Digital Forensic Education*,
Studies in Big Data 61, https://doi.org/10.1007/978-3-030-23547-5_2

1.2 Attorney's Questions[2]

The Attorney General needs answers to the following questions:

- At what time was the illegal drug lab raided?
- Could any of the two friends of Jessie Pinkman have been involved in the raid?

  - If yes:
    Which friend?
    What is the confidence in such hypothesis?

- How was the QBee camera disabled?

1.3 Quick Answers

1.3.1 At What Time Was the Illegal Drug Lab Raided?

According to the testimonies of the two associates, Varga and Pandana, Jessie Pinkman is said to be within the lab when the iSmartAlarm is set to "home" mode. At 10:34:17, "TheBoss" set the iSmartAlarm to "home" mode. Shortly after at 10:34:31, the iSmartAlarm was disarmed by a user named "pandadodu". This was followed by the smoke alarm alert at 10:36:06. Therefore, it is likely the raid occurred during this time frame.

1.3.2 Could Any of the Two Friends of Jessie Pinkman Have Been Involved in the Raid?

Three men were photographed inside the lab before the raid around 09:40:00. We are assuming these men are Pinkman, Pandana, and Varga. Additionally, the iSmart Alarm user "pandadodu" is likely the username for D. Pandana due to the similarity in the names. The pandadodu user disarmed the alarm and likely played a part in the attempted fire. These events give us the confidence to assume that Pandana was involved, but our confidence is limited by being unable to identify the men in the photographs, and so the assumption is not absolute.

1.3.3 How Was the QBee Camera Disabled?

We were unable to find information relating to this question.

---

[2]See footnote 1.

2 Case Reconstruction

The Digital forensic specialist observes some digital devices (Time Zone: UTC+2).

1. Physical extraction of Jessie Pinkman's Samsung phone
2. iSmartAlarm—Diagnostic logs
3. iSmartAlarm—Memory images
4. Arlo—Memory image
5. Arlo—NVRAM settings
6. NAND: TAR archive of the folder/tmp/media/nand
7. WinkHub—Filesystem TAR archive
8. Amazon Echo—Extraction of cloud data obtained via CIFT
9. Network capture.

2.1 Digital Crime Scene Reconstruction

The blue dotted line indicates devices that communicate via near field communication technology. Devices listed with a "?" are the assumed location of the device (Fig. 1).

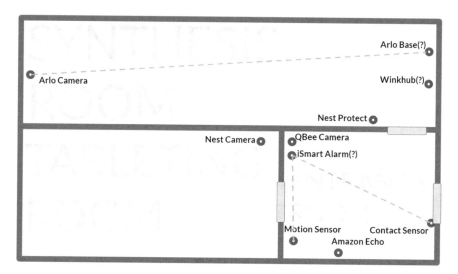

Fig. 1 Lab layout

2.2 Events/Timeline Reconstruction

Based on the data collected from the crime scene, we were able to create the following timeline. By listing the events in chronological order, this timeline may prove useful in the investigation (Fig. 2, Tables 1 and 2).

Visual Timeline

Key:

Green   iSmartAlarm
Pink   Door sensor
Blue   Motion sensor
Red   Nest protect
Yellow Police activity

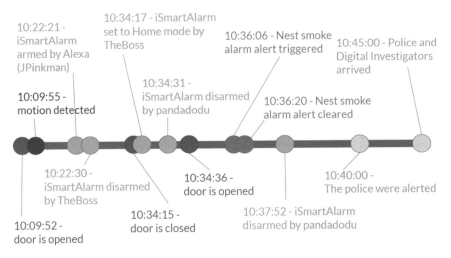

Fig. 2 Visual event timeline

Table 1 Event timeline source file location

Source label	File path
Amazon Echo	(2018-07-01_13.17.01)_CIFT_RESULT/cift_amazon_alexa.db
Server stream	diagnostics/2018-05-17T10_54_28/server_stream
Phone (iSmartAlarm)	/img_blk0_sda.bin/vol_vol21/data/iSA.common/databases/iSmartAlarm.DB
Phone (Nest Protect)	/img_blk0_sda.bin/vol_vol21/data/com.nest.android/cache/cache/cache-1332523362.json

Table 2 Event timeline

Time	Description	Source
May 17, 2018 09:44:53	Door sensor detects door is open	Server stream, Phone (iSmart Alarm)
May 17, 2018 09:45:22	iSmartAlarm disarmed by TheBoss	Server stream, Phone (iSmart Alarm)
May 17, 2018 09:47:18	Door sensor detects door is closed	Server stream, Phone (iSmart Alarm)
May 17, 2018 09:47:50	iSmartAlarm armed by JPinkman	Server stream, Phone (iSmart Alarm)
May 17, 2018 10:09:52	Door sensor detects door is open	Server stream, Phone (iSmart Alarm)
May 17, 2018 10:09:55	Motion Sensor detected motion	Server stream, Phone (iSmart Alarm)
May 17, 2018 10:09:57	iSmartAlarm disarmed by TheBoss	Server stream, Phone (iSmart Alarm)
May 17, 2018 10:22:22	iSmartAlarm armed by JPinkman via Alexa	Server stream, Phone (iSmart Alarm), Amazon Echo
May 17, 2018 10:22:30	iSmartAlarm disarmed by TheBoss	Server stream, Phone (iSmart Alarm)
May 17, 2018 10:34:15	Door sensor detects door is closed	Server stream, Phone (iSmart Alarm)
May 17, 2018 10:34:17	iSmartAlarm set to home mode by TheBoss	Server stream, Phone (iSmart Alarm)
May 17, 2018 10:34:31	iSmartAlarm was disarmed by pandadodu	Server stream, Phone (iSmart Alarm)
May 17, 2018 10:34:36	Door sensor detects door is open	Server stream, Phone (iSmart Alarm)
May 17, 2018 10:36:06	Nest Protect smoke alert triggered	Phone (Nest Protect)
May 17, 2018 10:36:20	Nest smoke alarm alert cleared	Phone (Nest Protect)
May 17, 2018 10:37:52	iSmartAlarm was disarmed by pandadodu	Server stream, Phone (iSmart Alarm)
May 17, 2018 10:40:00	The police were alerted	Challenge details
May 17, 2018 10:45:00	Police and digital investigators arrived	Challenge details

3 Data Analysis and Major Findings

3.1 Samsung Galaxy S6 Edge

We were able to find many application packages on the Samsung Phone. Below we have reported our major findings for each of the relevant applications (Table 3).

Table 3 Samsung Galaxy S6
edge file source

Physical extraction of Jessie Pinkman's Samsung phone	
File/Folder	Samsung GSM_SM-G925F Galaxy S6 Edge.7z
SHA256	ae83b8ec1d4338f6c4e0a312e73d7b410904fab50 4f7510723362efe6186b757

3.1.1   ISmartAlarm Application

Within the TB_IPUDairy table (pictured above), we were able to gather logged events for arming, disarming, and setting the iSmartAlarm to home mode. Additionally, these events indicated a user that performed the action. There were three users within the database—TheBoss, pandadodu, and JPinkman. By using the information within this table, we were able to narrow down the time window for the occurrence of the raid (Fig. 3, Table 4).

Varga and Pandana suggested that Jessie Pinkman would be within the lab when the iSmartAlarm was set to home mode. Assuming they were telling the truth, we considered the last home mode entry to be before the occurrence of the raid. After translating the epoch time and adjusting for timezone, the time for that event would

Table	TB_IPUDairy ⌄	30 entries	Page 1 of 1	← →	Export to CSV

▽ date	... IPUID	logType	... operator	... ...	profileid	profileName
1526546272	004D3209D9E4	2	pandadodu		2	DISARM
1526546071	004D3209D9E4	2	pandadodu		2	DISARM
1526546057	004D3209D9E4	2	TheBoss		1	HOME
1526545350	004D3209D9E4	2	TheBoss		2	DISARM
1526545342	004D3209D9E4	2	JPinkman		0	ARM
1526544597	004D3209D9E4	2	TheBoss		2	DISARM
1526543270	004D3209D9E4	2	JPinkman		0	ARM
1526543122	004D3209D9E4	2	TheBoss		2	DISARM

Fig. 3 iSmartAlarm TB_IPUDairy table

Table 4 iSmartAlarm application package location

File path	SHA1
/img_blk0_sda.bin/vol_vol21/ data/iSA.common/databases/iSmartAlarm.DB	37C924F86BF6E045CD1030E829D5984FC1 2865CE

be 10:34:17 May 17th. The next two entries are disarm events by the pandadodu user at 10:34:31 May 17th and 10:37:52 May 17th respectively. Due to the timing of the events, it is likely that the pandadodu user was involved in the raid. Additionally, the user pandadodu is very similar to Pandana's name. It may be that this user was intended for Pandana's use when accessing the lab (Fig. 4).

The TB_SensorDairy table listed sensor events. These events listed a sensor id and a date. Alone, this table is not that useful because there is not an accompanying table describing the sensor ids. However, with the iSmartAlarm server stream file provided by the digital investigators, we were able to identify which sensor id belonged to the motion sensor as well as which sensor id belonged to the door sensor found in the lab. These events added valuable information for understanding the timeline of the raid.

Table	TB_SensorDairy ⌄	41 entries	Page 1 of 1 ← →				Export to CSV
sensorID	▽ date	action	... model	operator	name	logtype	
000A8540	1526561895398	4	1	004D3209D9E4			
000A8540	1526561883694	3	1	004D3209D9E4			
000A8540	1526561826842	4	1	004D3209D9E4			
000A8540	1526561530075	3	1	004D3209D9E4			
000A8540	1526549990026	4	1	004D3209D9E4			
000A8540	1526546076051	3	1	004D3209D9E4			
004D3209D9E4	1526546057	0		700911	TheBoss	1	
000A8540	1526546055471	4	1	004D3209D9E4			
004D3209D9E4	1526545350	2		700911	TheBoss	1	
004D3209D9E4	1526544597	2		700911	TheBoss	1	
0006B4E5	1526544595279	5	1	004D3209D9E4			
000A8540	1526544592771	3	1	004D3209D9E4			
000A8540	1526543238967	4	1	004D3209D9E4			
004D3209D9E4	1526543122	2		700911	TheBoss	1	
000A8540	1526543093343	3	1	004D3209D9E4			

Fig. 4 iSmartAlarm TB_SensorDairy table

3.1.2 NestLabs Application

The NestLabs application for Android allows users to monitor and control their NestLab devices. In our case, the Nest Protect and Nest Camera (Fig. 5, Table 5).

Pictured above is the parsed json file using a free online json parsing tool: http://json.parser.online.fr/.

These are two key events recorded by the Nest Protect. The figure to the left is the event "protect_smoke_warn" time stamped 1526546166, and the figure to the right is the event "protect_smoke_warn_clear" time stamped 1526546180.

These events indicate that the Nest Protect smoke alarm went off at 10:36:06 on May 17, 2018, and the smoke alarm was subsequently cleared at 10:36:20 on May 17, 2018. The time stamps were converted from Unix Epoch time to UTC +2:00.

```
{
    "thread_id":1526546166676,
    "read":false,
    "priority":1,
    "timestamp":1526546166,
    "dismissed":true,
    "key":"protect_smoke_warn",
    "id":"577dda00-59ad-11e8-bce3-
    12f5f322bc9e",
    "parameters":[
        "18B430000039E345",
        "00000000-0000-0000-0000-
        00010000000a",
        "Kitchen",
        "LabSmoker",
        "SuperLab",
        2,
        1,
        1,
        2,
        1,
        1,
        "a3e36480-5757-11e8-80bb-
        0e2d565eed46",
        "Europe\/Zurich"
    ]
},
```

```
{
    "thread_id":1526546166676,
    "read":false,
    "priority":1,
    "timestamp":1526546180,
    "dismissed":true,
    "key":"protect_smoke_warn_clear",
    "id":"5fab5b80-59ad-11e8-bce3-
    12f5f322bc9e",
    "parameters":[
        "18B430000039E345",
        "00000000-0000-0000-0000-
        00010000000a",
        "Kitchen",
        "LabSmoker",
        "SuperLab",
        0,
        0,
        0,
        2,
        1,
        1,
        "a3e36480-5757-11e8-80bb-
        0e2d565eed46",
        "Europe\/Zurich"
    ]
}
```

Fig. 5 Nest protect Json

Table 5 Nest protect json file from NestLab application

File path	SHA1
/img_blk0_sda.bin/vol_vol21/data/com.nest.android/cache/cache/cache-1332523362.json	48c04e40187c9b5c2e10a2c30c98d3490ce039ed

Table 6 Google mail file location

mailstore.jpinkman2018@gmail.com.db	
File/Folder	/img_blk0_sda.bin/vol_vol21/data/com.google.android.gm/databases/ mailstore.jpinkman2018@gmail.com.db
SHA1	5B8AFED9A16BEBA35967B3F0E908979C98DADC1A

3.1.3 Google Mail

In the messages table, there is a column titled "bodyCompressed" that held Binary Large OBjects (BLOBs). Upon looking at the BLOBS in WinHex, we were able to identify the file signature "0x789C" which corresponds to zlib. zlib is a data compression algorithm. After decompressing these BLOBs, we found that this column in the database file contained the email body in html format (Table 6).

Using the above method, we were able to recover 40 emails. 20 of the emails were related to IoT devices. Those 20 emails contained information about Nest/iSmart Alarm alerts, Nest Aware subscription statuses, and a Nest safety summary report. These emails can be viewed at the following google drive link: https://drive.google. com/drive/folders/10yIsVTbXwZOeYTkLWVu2DBZaay-H_YsW?usp=sharing.

Unfortunately, none of the emails provided any valuable information for answering the Attorney General's questions.

3.2 ISmartAlarm

Our main goal for the server stream data was (Table 7):

- Confirm the findings found in the Samsung Phone
- Gain more information about sensor events
- Find additional sensor events not included in the Samsung Phone.

We were able to identify valuable information that logged system events within the file. This information can be found at the bytes 0xE003C–0x1377D6.

Below is a door sensor entry with highlighted sections.

From our investigation we have identified that (Fig. 6):

- Blue: Epoch Timestamp (UTC 0) in Hexadecimal
- Red: System Function

Table 7 iSmartAlarm server stream file location

iSmartAlarm—Diagnostic logs	
File/Folder	ismartalarm/diagnostics/2018-05-17T10_54_28/server_stream
SHA256	8033ba6d37ad7f8ba22587ae560c04dba703962ed16ede8c36a55c9553913736

```
$@00000005AFD3386: APSEND : 000A851C 0001 000107
$@00000005AFD3386: APSEND :the receive message is AP auto send, try to get more message
$@00000005AFD3386: ALARMDOOR : door is closed, and send to cloud
```

Fig. 6 iSmartAlarm door sensor entry

- Green: Details
- Purple: Sensor ID
- Orange: Sensor Event Counter

It should be noted that epoch timestamps are UTC 0. The time zone given in the challenge details was UTC+2. Therefore, any times found in the server stream must have two hours added in order to account for the time zone.

By correlating the events found in the Samsung Phone with the events logged in the server stream, we were able to add more detailed information than what was available in the Samsung Phone. Mainly, the server stream events enabled us to identify which sensor id belonged to each sensor. Additionally, as in the door sensor's case, we were able to identify whether the door sensor event indicated the door was opened or closed.

In the entrance of the lab, there was a motion sensor. Naturally, the iSmartAlarm would log the motion events in the server stream. As a result, we have aggregated the motion sensor data for May 17th, and we created a graph to visualize the motion events. Each dot indicates that motion was detected at that time (Fig. 7).

These motion sensor events further support the timeline of events for the raid. However, it should be noted that the motion sensor was only able to detect motion within the entrance of the lab. It should be emphasized that an absence of motion data does not mean there was absence of people in the lab, as someone could be in any of the other rooms of the lab during that time.

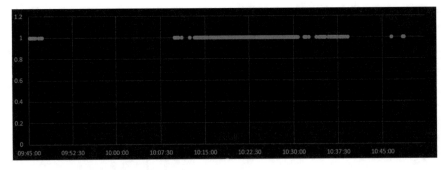

Fig. 7 iSmartAlarm motion sensor timeline

3.3   Amazon Echo

Pictured above is a snippet of the "ACCOUNT" table in the Alexa database file. This shows that the Echo is linked to Jessie Pinkman's email account, and that he is likely the owner of the device (Figs. 8 and 9).

The picture above is a snippet of the "TIMELINE" table in the Alexa database file. A person asked Alexa to arm the iSmartAlarm system and the mode was changed subsequently. After cross referencing this event with the iSmartAlarm database found on the phone, we deduced that the iSmartAlarm user JPinkman is associated with the Amazon Echo device (Tables 8 and 9).

	customer_email	customer_name
	Filter	Filter
1	jpinkman2018@gmail.com	Jessie Pinkman
2	jpinkman2018@gmail.com	Jessie Pinkman
3	NULL	Jessie Pinkman

Fig. 8 Amazon Echo user information

2018-05-17	10:22:20.720	UTC+2	Your system will set to Arm in 30 seconds.
2018-05-17	10:22:20.718	UTC+2	Mode Changed (iSmartAlarm)
2018-05-17	10:22:20.718	UTC+2	yes
2018-05-17	10:22:19.409	UTC+2	yes
2018-05-17	10:22:13.530	UTC+2	Your Door is open, Are you sure you want to arm your system?
2018-05-17	10:22:13.528	UTC+2	Mode Changed (iSmartAlarm)
2018-05-17	10:22:13.528	UTC+2	tell i. smart alarm to arm my system
2018-05-17	10:22:12.093	UTC+2	tell i. smart alarm to arm my system
2018-05-17	10:22:09.229	UTC+2	alexa

Fig. 9 Amazon Echo user commands timeline

Table 8 Amazon Echo database file information

Amazon Echo - cift_amazon_alexa.db	
File/Folder	(2018-07-01_13.17.01)_CIFT_RESULT/cift_amazon_alexa.db
SHA256	cc25acaac221db6fb17aaaa59409da94a7d178df126b72e7efaa7a301d85d0a6

Table 9 Network capture file source

Network capture	
File/Folder	network/dfrws_police.pcap
SHA256	1837ee390e060079fab1e17cafff88a1837610ef951153ddcb7cd85ad478228e

3.4 Network Capture

The picture above is the conversations window from the Statistics tab in Wireshark. The notable columns are: Address A (Source mac address), Address B (Destination mac address), Abs Start (Start time), and Duration (Fig. 10, Table 10).

Based on the pcap file and data found within the Samsung Phone, we were able to identify MAC addresses and local IPv4 addresses for devices found within the lab. Below is a table aggregating our findings. If there is supporting data from the phone, the file location of that information is listed. If there is not a supporting data from the phone *and* the MAC address does not appear in the pcap file, we are not entirely sure if this is the correct MAC address and have labeled the MAC address with a "?".

4 Additional Findings

Below are some photos we found in the Samsung Phone of the lab. They provide insight into who may be involved with the labs daily activities. We have divided the photos into two groupings—before the raid occurred and after the raid occurred.

Address A	Address B	Packe	Bytes	Packe	Bytes	Packe	Bytes	Abs Start	Duration
NestLabs_61:c9:ef	Raspberr_0e:3b:45	3,924	1593 k	2,132	1474 k	1,792	118 k	08:36:25.971725	57.0944
Raspberr_0e:3b:45	AskeyCom_e1:01:92	103	12 k	51	7532	52	5024	08:36:25.979027	56.2716
SamsungE_73:e3:78	Raspberr_0e:3b:45	117	16 k	56	6595	61	10 k	08:36:26.058577	49.8121
AmazonTe_96:23:24	Raspberr_0e:3b:45	34	4977	26	4115	8	862	08:36:27.744221	54.4226
Netgear_ff:75:4f	Raspberr_0e:3b:45	20	2030	10	1166	10	864	08:36:27.839911	45.0918
SamsungE_73:e3:78	AskeyCom_e1:01:92	12	1652	8	1236	4	416	08:36:34.874783	45.0061
SamsungE_25:02:fa	Raspberr_0e:3b:45	13	1368	6	643	7	725	08:36:45.870470	35.0425
NestLabs_99:9f:85	Raspberr_0e:3b:45	6	884	3	446	3	438	08:36:57.797023	6.6424
IPv4mcast_7f:ff:fa	Raspberr_0e:3b:45	20	10 k	0	0	20	10 k	08:37:01.000572	0.1101

Fig. 10 Wireshark conversations

Table 10 Device information

Device name	MAC address	IP address	In PCAP	Samsung file location
Arlo base station	08:02:8E:FF:75:4F	10.20.30.17	Y	/img_blk0_sda.bin/vol_vol21/media/0/DCIM/Camera/20180 326_164922.jpg
Wink Hub	B4:79:A7:25:02:FA	10.20.30.22	Y	/img_blk0_sda.bin/vol_vol21/media/0/DCIM/Camera/20180 410_091838.jpg
Samsung phone	AC:5F:3E:73:E3:78	10.20.30.21	Y	/img_blk0_sda.bin/vol_vol6/wifi/.mac.info
Qbee camera	D8:FB:5E:E1:01:92	10.20.30.15	Y	/img_blk0_sda.bin/vol_vol21/media/0/DCIM/Screenshots/Screenshot_20180 502-132904.png
Nest camera	18:B4:30:61:C9:EF	10.20.30.13	Y	/img_blk0_sda.bin/vol_vol21/media/0/DCIM/Camera/20180 410_092120.jpg
Nest protect	18:B4:30:99:9F:85	10.20.30.19	Y	/img_blk0_sda.bin/vol_vol21/media/0/DCIM/Camera/20180 410_091924.jpg
Amazon Echo	74:75:48:96:23:24	10.20.30.23	Y	
Router (Raspberry pi)	B8:27:EB:0E:3B:45	10.20.30.1	Y	
iSmart Alarm	00:4D:32:09:D9:E4			/img_blk0_sda.bin/vol_vol21/media/0/DCIM/Camera/20180 410_092015.jpg

Note: The times for the images below are the timestamps given from what we believe is the time they were downloaded to the phone, so we do not know exactly when these pictures were taken.

4.1 Before the Raid Occurred at 9:39–9:40 May 17th

We believe the police may be able to identify the individuals in these photos. We believe these individuals may be Pinkman, Varga and Pandana. This may suggest that Pandana and Varga may be more involved in the labs daily operations than what they told the police (Table 11, Fig. 11).

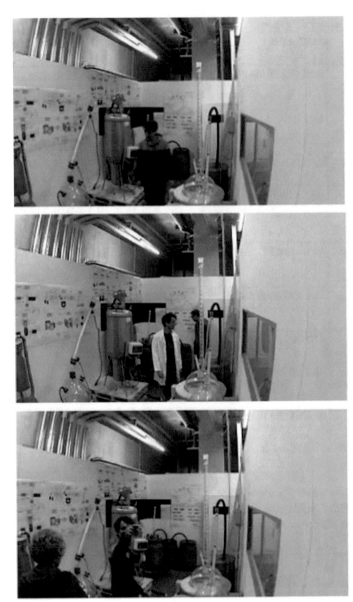

Fig. 11 4–3 photos before the raid

Table 11 Photos before the raid

Pictures	
Location on Samsung phone	/img_blk0_sda.bin/vol_vol21/data/com.netgear.android/cache/http
File names	1. 0860f6c5ed0c20de5694f7efd37b94c7.0 2. fd6f9b0229627dbc749f065b67e0e72b.0 3. 67f43186731b8ca1b0dbf25ec25c5391.0
MD5	1. 870870fcabed361d471c6c6fc10dbb7c 2. ea93548a5f0665e3eb2e10ba63d98201 3. 5478b3842db1c1944ce07fe4d231d762

4.2 After the Raid Occurred at 19:32 May 17th

We are unsure as to who these individuals could be. Due to the time, it may possible that these are the investigators on scene (Table 12, Fig. 12).

Table 12 Photos after the raid

Pictures	
Location on Samsung phone	/img_blk0_sda.bin/vol_vol21/data/com.netgear.android/cache/http
File names	1. 3003e82dd20d57e22ed85d9679acf041.0 2. e25fb6ddac14ba4eedae779a679cd129.0 3. 9740fe57972a6de0498ee39560ca2f1c.0
MD5	1. 20e45fdcb1468f1e5442634c59e54998 2. 88ac0e33fb0e2b4c2bf08ffc4d988f23 3. 227e9457243d029cb0b7864f46dc5e0b

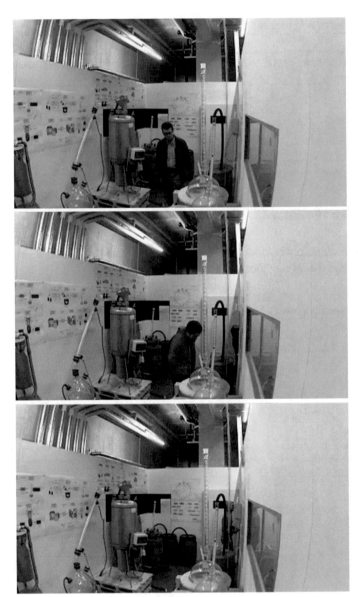

Fig. 12 4–6 photos after the raid

DFRWS IoT Forensic Challenge Report 2

Jose Reyna, Madison Evans Larrumbide and Carson Bosko

Abstract In this chapter, we report on the findings of our group's analysis of the Digital Forensic Research Workshop (DFRWS) IoT Forensic Challenge (2018–2019).

1 Case Summery

1.1 Case Breakdown[1]

- 10:40 05-17-2018
- The police received an alert that an illegal drug lab was set on fire.
- 10:45 2018-05-17
- Police and a digital forensic specialist arrive on scene. Jessie Pinkman, the owner of the illegal drug lab, is missing.

The owner the illegal drug lab, Jessie Pinkman, is nowhere to be found. Police interrogate two of Jessie Pinkman's known associates: D. Pandana and S. Varga. Pandana and Verga admit having access to the drug lab's WiFi network but deny any involvement in the raid. They also say that Jessie Pinkman's had the IoT security systems installed because he feared attacks from a rival gang and that Jessie kept the alarm engaged in "Home" mode whenever he was inside the drug lab.

Within the drug lab the digital forensic specialist observes some IoT devices, including an alarm system (iSmartAlarm), three cameras (QBee Camera, Nest Camera and Arlo Pro) as well as a smoke detector (Nest Protect). An Amazon Echo and a WinkHub are also present.

---

[1] https://www.dfrws.org/dfrws-forensic-challenge.

J. Reyna (✉) · M. E. Larrumbide · C. Bosko
Department of Information System and Cyber Security, University
of Texas at San Antonio, San Antonio, TX, USA
e-mail: j.mreyna@yahoo.com

© Springer Nature Switzerland AG 2020
X. Zhang and K.-K. R. Choo (eds.), *Digital Forensic Education*,
Studies in Big Data 61, https://doi.org/10.1007/978-3-030-23547-5_3

1.2 Attorney's Questions[2]

The Attorney General needs answers to the following questions:

1. At what time was the illegal drug lab raided?
2. Could any of the two friends of Jessie Pinkman have been involved in the raid?
3. If yes, Which friend?
4. What is the confidence in such hypothesis?
5. How was the Qbee camera disabled?

1.3 Quick Answers

1. At what time was the illegal drug lab raided?

   a. Given that the smoke alarm is set off at 10:36 A.M.—we assume a direct result of the raid, and the police are called four minutes later, the raid will have taken place within a reasonable amount of time after 10:34:30, the time that Pandana disables the alarm system.

2. Could any of the two friends of Jessie Pinkman have been involved in the raid?

   a. Yes, Pandana is believed to the involved in the raid, given that we place the raid at 10:34:30 to 10:36:00, Pandana disabling the alarm at 10:34:31 indicates that he had an involvement.

3. If yes, Which friend?

   a. D. Pandana (see above)

4. What is the confidence in such hypothesis?

   a. The chain of events and audit/log files that we have that place users at these events at exact times give us a strong indication of the timeline and crime. As such, our hypothesis is backed by strong confidence.

5. How was the Qbee camera disabled?

   a. While we still do not know how or why the QBee was disabled, we assume it had something to do with the raid. Perhaps Pandana was involved in the disconnecting of the QBee—while this assumption makes sense, we have no evidence to back it up and as such submit that we do not know how the camera was disabled (Fig. 1).

---

[2]https://www.dfrws.org/dfrws-forensic-challenge.

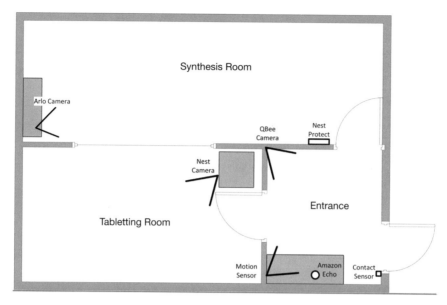

Fig. 1 Physical layout of lab (provided by law enforcement)

2 Reconstruction

2.1 Crime Scene Reconstruction

Observing the PCAP file that was provided by law enforcement, we built out the addressing information for the devices found at the crim scene (See Networking PCAP analysis for details).

The list of all physical devices is shown in Table 1. The IP addresses were found in the PCAP file. Pictures of the devices 2 through 8 were found on the Samsung Phone image showed the MAC addresses. The MACs for Devices 9 and 10 were found in communication log files on the phone image. Not all the devices are critical to our findings (Device names with a "*" denote an important device in our investigation:

The virtual layout of the crime scene is shown in Fig. 2. The mapping of device communication is cross referenced by the network PCAP file provided. The communication between devices is represented by arrows (sometimes bi-directional communication). Almost all devices talk directly to the router. The Arlo Pro camera communicates via the Arlo Base Station (a hub for Arlo devices). The Door Sensor and Motion Sensor both communicate with the iSmart Alarm to provide an alarm state. The understanding of the Virtual layout and the physical layout are critical to understanding what devices might hold pertinent data and information to the investigation. Since this is an IoT investigation, the data necessary for the investigation could be on any device, so understanding possible location of data is important. Note:

Table 1. The IP and MAC address for the digital devices found from the crime scene

Device	IP address	Mac address	Image available
1. Raspberry Pi Router	10.20.30.1	B8:27:EB:0E:3B:45	No
2. Nest Camera	10.20.30.13	18:B4:30:61:C9:EF	No
3. QBee Camera	10.20.30.15	D8:FB:5E:E1:01:92	No
4. Arlo Base Station	10.20.30.17	08:02:8E:FF:75:4F	Yes
5. iSmartAlarm*	10.20.30.18	00:4D:32:09:D9:E4	Yes
6. Nest Smoke Alarm*	10.20.30.19	18:B4:30:99:9F:85	No
7. Samsung Galaxy Edge 6*	10.20.30.21	AC:5F:3E:73:E3:78	Yes
8. WinkHub	10.20.30.22	B4:79:A7:25:02:FA	Yes
9. Amazon Echo*	10.20.30.23	74:75:48:96:23:24	Yes
10. Arlo Camera	via Arlo Base station	00:90:4C:1E:00:01	Yes

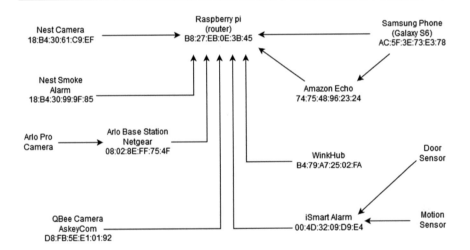

Fig. 2 Virtual layout of crime scene

The phone is a central device because it is a facilitator of communication between multiple devices, as such, much of our findings and focus was on the phone image.

Of the devices found at the scene, the **Samsung Phone, Amazon Echo, Nest Smoke Alarm**, and **iSmart Alarm** are the most important to our investigation. All four devices had critical evidence that built the timeline. The phone provided images and communication information between the alarm system and physical sensors. The iSmart Alarm was key in the reconstruction of the virtual layout and timeline as well as understanding events. The Echo and Smoke Alarm evidence was important in understanding when the raid took place.

2.2 Timeline and Events

The timeline below is built using the following assumptions that we are confident about and is supported by the forensic evidence (Sections for more details listed).

- The loud beep in the sound file located on the Amazon Echo is the alarm system going off because the door is open while the system is armed. (See Amazon Echo Analysis)
- Pandana arrives at 10:34, this is his first interaction with the alarm system. (See iSmart Alarm Analysis)
- The smoke that is detected from the Nest Smoke Alarm at 10:36 is a direct result of the raid. (See Nest Smoke Alarm Analysis)
- The QBee camera is disconnected prior to the raid., evidenced by the repeated and unsuccessful pinging of the QBee camera after this time. (See PCAP Analysis)
- The door sensor is audited as "TheBouncer" (See iSmart Alarm Analysis)
- The motion sensor is audited as "TheMotion" (See iSmart Alarm Analysis)
- Alarm voice commands issued via the Amazon Echo are audited as "TheBoss" (See iSmart Alarm Analysis).

Timeline

- 10:09:52 door is opened by Pinkman[3]

    – Pinkman Arrives

- 10:22:21—Alarm armed by Pinkman[4]
- 10:22:30—Alarm goes off because door is open[5]
- 10:34:15—Door closed[6]
- 10:34:17—The Boss sets alarm to home mode[7]
- 10:34:31—Alarm system is disarmed by Pandana[8]
- 10:34:36—Door opened[9]

    – Pandana arrives

- 10:34–10:37—Raid occurs
- 10:36:06—Nest Smoke Alarm alerts starts[10]
- 10:36:20—Nest Smoke Alarm alert stops[11]

---

[3] See iSmartAlarm analysis for details.

[4] See iSmartAlarm analysis for details.

[5] See Amazon Echo analysis for details.

[6] See iSmartAlarm analysis for details

[7] See iSmartAlarm analysis for details.

[8] See iSmartAlarm analysis for details.

[9] See iSmartAlarm analysis for details.

[10] See Nest Smoke Alarm analysis for details.

[11] See Nest Smoke Alarm analysis for details.

- 10:37:52—Pandana disarmed alarm[12]
- 10:40:00—Police are called
- 10:45:00—Police arrive.

3 Data Analysis

3.1 Important Devices

We found four devices to the most important in our investigation. They are listed here as Is their main value in reconstruction. Details of where the information was located and in-depth analysis as to how it pertains to the case is listed in their detailed sections.

Samsung Edge S6 Phone

- Most of the evidence was found on the phone image.

Amazon Echo

- Sound file with alarm beeping at 10:22.

iSmart Alarm

- Alarm status (Arm/Disarm audit log).

Nest Smoke Detector

- Smoke detection at 10:36 A.M.

3.2 Network PCAP of Raspberry Pi Raouter

A capture of the network traffic was taken as part of the initial crime scene investigation. The Network PCAP file was provided by the police forensics investigators. We opened it in Wireshark and used it to identify all the devices that were active at the time of the crime. Over the course of the investigation the PCAP was used for cross-referencing and validating communication evidence that we had seen.

Using Wireshark early in the process, we found the following addressing information for the devices. (The devices are listed in order they appear on Wireshark) (Separated by device clarity) (Figs. 3, 4, 5, 6, 7, 8, and 9).

The PCAP is the source of a major point of evidence for our timeline. The QBee is disconnected at the time of the capture, the PCAP showing repeated pings to the device without response, this led us to believe the QBee had been disconnected prior to the raid.

---

[12] See iSmartAlarm analysis for details.

```
178 13:36:27.791433   Raspberr_0e:3b:45   SamsungE_73:e3:78   ARP   42 Who has 10.20.30.21? Tell 10.20.30.1
181 13:36:27.825669   SamsungE_73:e3:78   Raspberr_0e:3b:45   ARP   42 10.20.30.21 is at ac:5f:3e:73:e3:78
```

Fig. 3 Samsung Galaxy S6 ARP

```
533 13:36:32.875193   Netgear_ff:75:4f    Raspberr_0e:3b:45   ARP   60 Who has 10.20.30.1? Tell 10.20.30.17
534 13:36:32.875218   Raspberr_0e:3b:45   Netgear_ff:75:4f    ARP   42 10.20.30.1 is at b8:27:eb:0e:3b:45
```

Fig. 4 Raspberry Pi router ARP

```
1817 13:36:50.191449  Raspberr_0e:3b:45   NestLabs_61:c9:ef   ARP   42 Who has 10.20.30.13? Tell 10.20.30.1
1818 13:36:50.193093  NestLabs_61:c9:ef   Raspberr_0e:3b:45   ARP   42 10.20.30.13 is at 18:b4:30:61:c9:ef
```

Fig. 5 Nest Camera ARP

```
1986 13:36:52.911434  Raspberr_0e:3b:45   Netgear_ff:75:4f    ARP   42 Who has 10.20.30.17? Tell 10.20.30.1
1988 13:36:52.920643  Netgear_ff:75:4f    Raspberr_0e:3b:45   ARP   60 10.20.30.17 is at 08:02:8e:ff:75:4f
```

Fig. 6 Arlo base station ARP

3.3 Samsung (S6 Phone)

Pinkman's phone image was the first major source of evidence we noted during our investigation. The evidence found on the phone pertained to multiple devices and avenues of investigation; images, passwords, emails, and encryption keys to name a few.

The images that head the sections on devices below were found on the phone image and fill in any gaps as to the MAC or IP addresses we needed for devices following the initial PCAP analysis.

Our knowledge as to the state of the alarm was found within the phone's databases. See the iSmart alarm section for more details on the alarm detection time.

Pinkman's email address was found in the phone—this email was helpful when tracing and users of devices. Found in the a database file of accounts, the Account_name column of **jpinkman2018 @gmail.com**, account type of com.google (Found at: **/img_blk0_sda.bin/ vol_vol21/data/com.android.providers.contacts/databases/contacts02.DB**).

The mac address for the phone itself was found at **/img_blk0_sda.bin/ vol_vol6/wifi/.mac.info,** giving us the mac address: **AC:5F:3E:73:E3:78**.

```
2725 13:37:02.831434  Raspberr_0e:3b:45   NestLabs_99:9f:85   ARP   42 Who has 10.20.30.19? Tell 10.20.30.1
2726 13:37:02.833086  NestLabs_99:9f:85   Raspberr_0e:3b:45   ARP   42 10.20.30.19 is at 18:b4:30:99:9f:85
```

Fig. 7 Nest smoke alarm ARP

```
3607 13:37:14.351404  Raspberr_0e:3b:45   AmazonTe_96:23:24   ARP   42 Who has 10.20.30.23? Tell 10.20.30.1
3608 13:37:14.359720  AmazonTe_96:23:24   Raspberr_0e:3b:45   ARP   42 10.20.30.23 is at 74:75:48:96:23:24
```

Fig. 8 Amazon echo ARP

```
4087 13:37:20.911446   Raspberr_0e:3b:45   SamsungE_25:02:fa   ARP        42 Who has 10.20.30.22? Tell 10.20.30.1
4088 13:37:20.912996   SamsungE_25:02:fa   Raspberr_0e:3b:45   ARP        42 10.20.30.22 is at b4:79:a7:25:02:fa
```

Fig. 9 WinkHub ARP

3.4 Amazon Echo

The Amazon Echo did not initially provide evidence as to the state of other devices, however it did provide us with a key piece of evidence that aided in the timeline reconstruction. There is a sound capture on the echo found at:

(2018-07-01_13.17.01)_CIFT_RESULT/Evidence_Library/AmazonAlexa Cloud/VOICE/(2018-05-17T10_22_24+0200)_TEXT(TRANSCRIPT NOT AVAILABLE).wav.

The sound file was initially thought to be the fire alarm going off—a direct result of the raid. However, after finding information in the nest smoke alarm that states that the smoke alarm alerts to the presence of smoke at 10:36, the loud beeping heard in the recording above is now thought to be the alarm system triggering because the door was left open while the system was armed. This assumption is further backed up by the iSmart Alarm audit logs of alarm state, the open door is indicated there (See iSmart Alarm section—Fig. 11).

3.5 ISmart Alarm

The MAC address of the alarm was found on an image (right) on the phone image at: located at **/img_blk0_sda.bin/vol_vol21/media/0/DCIM/Cam era/20180410_092015.jpg.** A critical part of the reconstruction, the data revolving the state of the alarm and the connected door and motion sensors was found mostly on the phone image. A lot of the timeline recreation was cross-referenced against the following log file found on the phone. The log file keeps track of the state of the alarm and is an audit log for requests to change the state of the alarm. (Found at **/img_blk0_sda.bin/ vol_vol21/data/iSA.common/databases/iSmartAlarm.DB**) (Figs. 10 and 11).

This log is a primary reason we believe that Pandana is involved in the raid, his entrance time in accordance with the time the nest smoke alarm detects smoke leads us to this conclusion (See Nest Smoke Alarm analysis).

On the phone we found a screenshot that helped us identify the names of the sensors in the audit log. The door sensor is called "TheBouncer", the motion sensor, "TheMotion", and "TheBoss" appears to be the name attributed to commands coming from the Amazon Echo. (this screenshot was found at (**/img_blk0_sda.bin/vol_vol21/media/0/DCIM/Screenshots/ Screenshot_20180330-203515.png**) on the phone image).

Fig. 10 iSmart alarm MAC

	date	action	IPUID	logType	sensorName	operator	sensorType	sensorID	userID	profileid	profileName
	Filter	Filter	Filter	Filter	Filter	Filter	Filter	Filter	Filter	Filter	Filter
1	2018-05-17 08:37:52		004D3209D9E4	2		pandadodu				2	DISARM
2	2018-05-17 08:34:31		004D3209D9E4	2		pandadodu				2	DISARM
3	2018-05-17 08:34:17		004D3209D9E4	2		TheBoss				1	HOME
4	2018-05-17 08:22:30		004D3209D9E4	2		TheBoss				2	DISARM
5	2018-05-17 08:22:22		004D3209D9E4	2		JPinkman				0	ARM
6	2018-05-17 08:09:57		004D3209D9E4	2		TheBoss				2	DISARM
7	2018-05-17 07:47:50		004D3209D9E4	2		JPinkman				0	ARM
8	2018-05-17 07:45:22		004D3209D9E4	2		TheBoss				2	DISARM
9	2018-05-16 13:55:27		004D3209D9E4	2		JPinkman				0	ARM
10	2018-05-16 13:55:17		004D3209D9E4	2		JPinkman				2	DISARM
11	2018-05-16 13:55:11		004D3209D9E4	2		JPinkman				2	DISARM
12	2018-05-16 13:55:07		004D3209D9E4	2		JPinkman				2	DISARM
13	2018-05-16 13:53:31		004D3209D9E4	2		JPinkman				0	ARM
14	2018-05-16 13:53:10		004D3209D9E4	2		JPinkman				1	HOME
15	2018-05-16 13:47:40		004D3209D9E4	2		TheBoss				2	DISARM
16	2018-05-16 13:47:33		004D3209D9E4	2		TheBoss				2	DISARM
17	2018-05-16 13:47:18		004D3209D9E4	2		TheBoss				1	HOME
18	2018-05-16 13:47:15		004D3209D9E4	2		TheBoss				3	PANIC
19	2018-05-16 13:47:14		004D3209D9E4	2		TheBoss				3	PANIC
20	2018-05-15 13:05:27		004D3209D9E4	2		JPinkman				2	DISARM
21	2018-05-15 13:04:43		004D3209D9E4	2		pandadodu				1	HOME
22	2018-05-15 13:03:47		004D3209D9E4	2		TheBoss				2	DISARM
23	2018-05-15 13:03:43		004D3209D9E4	2		pandadodu				3	PANIC
24	2018-05-15 13:03:36		004D3209D9E4	2		pandadodu				2	DISARM
25	2018-05-15 13:03:28		004D3209D9E4	2		pandadodu				3	PANIC
26	2018-05-15 13:03:11		004D3209D9E4	2		pandadodu				3	PANIC
27	2018-05-15 12:41:35		004D3209D9E4	2		JPinkman				2	DISARM
28	2018-05-15 12:38:20		004D3209D9E4	2		JPinkman				0	ARM
29	2018-05-15 09:38:53	2	004D3209D9E4	5							
30	2018-05-15 09:34:20	1	004D3209D9E4	5							

Fig. 11 Alarm Arm/Disarm audit log

Fig. 12 Nest smoke alarm
MAC and device
information. Found at
/img_blk0_sda.bin/
vol_vol21/media/0/DCIM/Cam
era/20180410_091924.jpg
on phone image

3.6 Nest Smoke Alarm

The nest Smoke Alarm is the last of the four most important devices
we found. The smoke alarm ties together our timeline and gives us evidence for the time of the raid. The smoke alarm goes off at 10:36 A.M.
The Nest Smoke alarm logs were helpful when understanding the time
when the raid took place. The logs were found at: **/img_blk0_sda.bin/
vol_vol21/data/com.nest.android/cache/cache/cache-1332523362.json**.

The information in the json file show that the smoke alarm detected smoke at
10:36:06 A.M. We consider the smoke alert to be a direct result of the raid. The
smoke detector stops alerting at 10:36:20. This, in conjunction with the police being
called at 10:40:00 lead us to conclusion about the time of occurrence for the raid and
the failed arson attempt (Fig. 12).

3.7 Arlo Camera

The Arlo camera connects via the base station, though it does not connect directly,
the phone caches images that are sent from the Arlo Camera. On the Samsung phone
image, there is cached pictures detailing the drug lab. (Found at: **/img_blk0_sda.bin/**

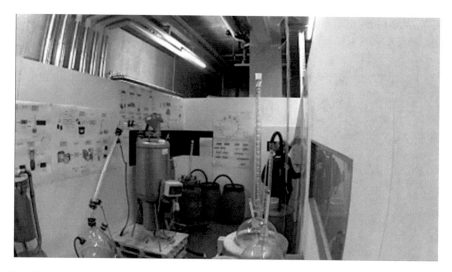

Fig. 13 Arlo Camera Picture 1 of Drug lab

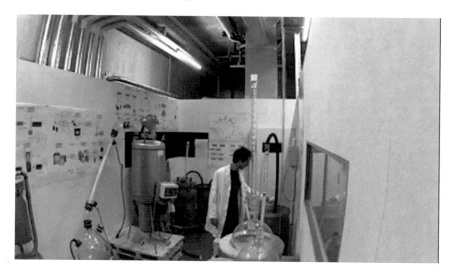

Fig. 14 Arlo Camera Picture 2 of Drug lab

vol_vol21/data/com/.netgear.android/cache/http) The pictures show four distinct figures that could be identified and the timeline leading up to the raid (Figs. 13, 14, 15 and 16).

Even though the Arlo images don't provide us with an additional idea of what happened during the raid they do help us understand the movements in the Synthesis room and the time leading up to the raid.

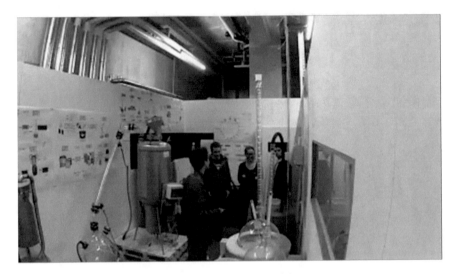

Fig. 15 Arlo Camera Picture 3 of Drug lab

Fig. 16 Arlo Camera Picture 4 of Drug lab

3.8 QBee Camera

Labeled as ASKey_Com in the PCAP, communications to the QBee camera are limited. However, we do know that it was disabled during the PCAP's capture, evidenced in the PCAP, the QBee (Ip address: 10.20.30.15; MAC address: D8:FB:5E:E1:01:92 (see Fig. 17)), is unreachable with no response. We have not been able to locate information on the QBee camera as to what images it may have captured before being

Fig. 17 QBee Camera MAC
and technical Information.
Found at /img_blk0_sda.bin/
vol_vol21/media/0/DCIM/Screenshots/
Screenshot_20180502-
132904.png on phone
image

disconnected. Our main assumption about the QBee Camera is that it was disabled deliberately and not by accident or electrical fault—the fact that it points at the drug lab's entrance and that it was disconnected prior to the raid lend credence to the idea that one of Pinkman's associates was involved in the raid, however this assumption alone does not imply which one of the associates it may be.

4 Assumptions and Other Findings

4.1 Assumptions

The timeline leading up to the raid could be built up more if positive identifications could be made on the individuals in the pictures found from the Arlo Caemra.

One of the audit log entries displays, Pandadodu, we assume this to be D. Pandana's username, as such we have him entering the lab at 10:34:36 A.M.

Pinkman putting the alarm system into "Home" mode and pandana's disabling of the alarm system around the same time indicate Pandana's involvement in the raid.

4.2 Other Findings

There is a database file on the phone image that has creation timestamps for accounts used, or when they were added. Using a timestamp decoding tool—we used DCode—we were able to determine that the account was created on 2016. (Found at: /img_blk0_sda.bin/vol_vol21/system/users/0/accounts.db on the phone image).

Public and private keys were found on the phone image, they may be able to be used to decrypt files on the phone could we identify them. (Found at: /img_blk0_sda.bin/vol_vol21/data/com.google.android.gms/databases/cryptauthkeys.db on phone image).

4.3 Observations

Most of the information on the phone prior to 5/15/2018 indicate that the phone set up including time was set to Zurich time zone. Which would be the original owner data to include road maps from Germany.

There are also indications found within the phones of many of the accounts that were used for the devices that were also set up a short period prior to challenge "offense date" which would also interfere with proper forensics based off time stamps.

However, these did not impede our investigation, they were interesting to note—accidental or otherwise.

DFRWS IoT Forensic Challenge Report 3

Paul Harris, James Ma, Ian Salas and Isaac Sanchez

Abstract In this chapter, we report on the findings of our group's analysis of the Digital Forensic Research Workshop (DFRWS) IoT Forensic Challenge (2018–2019).

1 Introduction

This semester our Final Project was to conduct a digital forensic investigation on IoT devices. IoT refers to the Internet of Things and refers to devices that are connected to the internet or a network. These devices drive the technology field in today's market. These devices are a part of our everyday lives, ranging from "Smart" Phones, "Smart" Watches, "Smart" TV's, and even "Smart" Cars. This new technology is the driving force within the technology market.

The purpose of this challenge is to learn and understood the complexities about IoT devices. It is to learn about developing techniques to acquire, investigate and interpret information that is crucial for a digital forensic investigation.

1.1 Challenge Scenario

Police are called to a drug lab suspected of belonging to Jessie Pinkman.

1.1.1 Details of Scenario

- 10:40 2018-05-17
  Police were called to a drug lab that was suspected of being invaded and subsequently set on fire

P. Harris · J. Ma (✉) · I. Salas · I. Sanchez
Department of Information System and Cyber Security,
University of Texas at San Antonio, San Antonio, TX, USA
e-mail: j.ma4227@gmail.com

© Springer Nature Switzerland AG 2020
X. Zhang and K.-K. R. Choo (eds.), *Digital Forensic Education*,
Studies in Big Data 61, https://doi.org/10.1007/978-3-030-23547-5_4

- 10:45 2018-05-17
  Forensic team arrived on scene at collected evidence
  Suspect "Jessie Pinkman" was not at location, but 2 known associates D. Pandana and S. Varaga has full access to the lab's wifi network
  Pinkman installed numerous IoT security devices in fear of a rival attack.
  Both Associates deny involvement with the invasion at the lab.

1.2 Challenge Questions

- At what time was the illegal drug lab raided?
- Could any of the two friends of Jessie Pinkman have been involved in the raid?

  – Which Friend?

- How was the Qbee camera disabled?

1.3 Device Diagram

Figure 1-1 is a diagram of the drug lab supplied by the DFRWS challenge. The diagram depicts the layout of the drug lab and the locations of each IoT device.

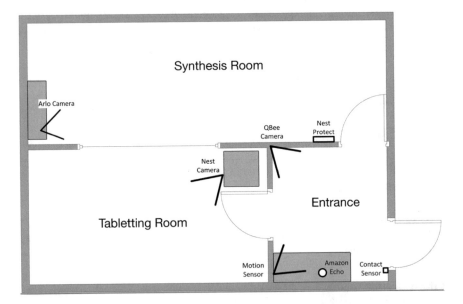

FIGURE SEQ FIGURE * ARABIC 1-DRUG LAB LAYOUT *IMAGE PROVIDED BY* DFRWS CHALLENGE

2 Observations and Findings

2.1 What We Know

- 3 individuals were shown on the Arlo cache images at around 2018-05-17 07:32 UTC
- iSmartAlarm log showed 3 users who accessed the lab between the hours of 07:45 UTC and 08:37 UTC
- Account names of "TheBoss", "JPinkman", "Panadodu"
- Pinkman arms the security system @ 08:22:22 UTC
- 'TheBoss' disarms the security system @ 08:22:30 UTC
- 'TheBoss' puts the system in Home mode @ 08:34:17 UTC
- iSmartAlarm showed status of "Home" @ 08:34:17 UTC
- Panadodu disarmed iSmartAlarm @ 08:34:31 UTC
- QBee Camera was unreachable @ IP 10.20.30.15 @ 08:36:14 UTC
- iSmartAlarm was disabled by user Panadodu at 2018-05-17 08:37 UTC
- Image from Arlo Cache shows individual possibly tampering with Qbee Camera @ Possibly 08:39 UTC
- 5 individuals were shown on the Arlo cache images at around 2018-05-17 17:32 UTC.

2.2 Findings (Challenge Questions Answered)

2.2.1 At What Time Was the Illegal Drub Lab Raided?

Based on our observations and evidence that was analyzed we can say that the raid may have happened between 08:34 UTC and 08:37 UTC.

2.2.2 Could Any of the Two Friends of Jessie Pinkman Have Been Involved in the Raid?

Based on the evidence collected and analysis of the evidence we can say that Pandana was involved in the raid.

The fact that Pandana was the last individual who was indicated on the iSmartAlarm log to have disabled the alarm there could be further evidence to indicate Pandana's direct involvement with the invasion of the drug lab.

2.2.3 How Was the Qbee Camera Disabled

The network capture file provided by the police indicated that there were unsuccessful UDP requests to the IP of 144.76.81.240 from the IP of 10.20.30.15. We already know that the IP address of the Qbee camera is 10.20.30.15. The data from the IP of 144.76.81.240 indicated an attempted connection from the application package of the Qbee Camera that was on the Samsugn Galaxy S6 Edge. Meaning that someone was attempting to connect to the camera remotely. The unsuccessful requests were at around 08:36:14 UTC. Meaning that the Qbee camera was unplugged and had no connection to the network.

3 Overview of Challenge Data

Evidence Data is located at
https://drive.google.com/drive/folders/17szABuOp3wEI9FAWsx5Q4ThAtjPecm4Q?usp=sharing
Time Zone: UTC + 2
During the collection of evidence the following devices were found:

1. iSmart Alarm—Memory Images

   - dump/ismart_00.img, b175f98ddb8c79e5a1e7db84eeaa691991939065ae17bad84cdbd915f65d9a10
   - dump/ismart_80.img, b175f98ddb8c79e5a1e7db84eeaa691991939065ae17bad84cdbd915f65d9a10

2. iSmartAlarm—Diagnostic logs

   - File/Folder: ismartalarm/diagnostics/2018-05-17T10_54_28/server_stream
   - SHA256: 8033ba6d37ad7f8ba22587ae560c04dba703962ed16ede8c36a55c9553913736

3. Arlo—Memory Image

   - File/Folder: arlo/dfrws_arlo.img
   - SHA256: 3b957a90a57e5e4485aa78d79c9a04270a2ae93f503165c2a0204de918d7ac70

4. Arlo—NVRAM settings

   - File/Folder: arlo/nvram.log
   - SHA256: 3b957a90a57e5e4485aa78d79c9a04270a2ae93f503165c2a0204de918d7ac70

5. Arlo—NAND: TAR archive of the folder/tmp/media/nand

   - File/Folder: arlo/arlo_nand.tar.gz

- SHA256: 857455859086cd6face6115e72cb1c63d2befe11db92beec52d1f706 18c5e421

6. Amazon Echo

   - File/Folder: echo/(2018-07-01_13.17.01)_CIFT_RESULT.zip
   - SHA256:   7ee2d77a3297bb7ea4030444be6e0e150a272b3302d4f68453e8 cfa11ef3241f

7. Winkhub—Filesystem TAR archive

   - File/Folder: wink/wink.tar.gz
   - SHA256: 083e7428dc1d0ca335bbcfc11c6263720ab8145ffc637954a7733afc 7b23e8c6

8. Network Capture

   - File/Folder:network/dfrws_police.pcap
   - SHA256: 1837ee390e060079fab1e17cafff88a1837610ef951153ddcb7cd85ad 478228e

9. Physical extraction of Jessie Pinkman's Samsung Phone

   - File/Folder: Samsung GSM_SM-G925F Galaxy S6 Edge.7z
   - SHA256: ae83b8ec1d4338f6c4e0a312e73d7b410904fab504f7510723362efe 6186b757

3.1   Device Descriptions

3.1.1   ISmart Alarm

iSmartAlarm[1] is a do-it-yourself smart home security system controlled with a user's smartphone. The system and devices are designed and manufactured by iSmart Alarm Inc. The system uses a hub connected to a router to allow users control of home security and home automation devices. Users can arm and disarm their system, notifications, emails and phone calls if the system is triggered. The iSmart Alarm is a self-monitored solution and is not monitored by a third party. It is also integrates with Amazon Alexa.

3.1.2   Arlo Pro Camera

Netgear Arlo is a smart home security camera. Device was collected during the physical acquisition of evidence at the drug lab. The Arlo Camera is positioned

---

[1]https://en.wikipedia.org/wiki/ISmartAlarm.

directly above the synthesis room and may provide a key piece of evidence upon analysis.

3.1.3   Amazon Echo

Amazon Echo is a smart speaker. The Echo responds to voice commands and can be integrated with most applications and devices such as smartphones, thermostats, home alarm systems etc. "Alexa" is the intelligent personal assistant that will respond to your requests.

```
●  ●  ●              DFRWS — -bash — 80×24
[Jamess-MacBook-Pro-2:DFRWS jamesma$ shasum -a 256 "(2018-07-01_13.17.01)_CIFT_RE]
SULT.zip"
7ee2d77a3297bb7ea4030444be6e0e150a272b3302d4f68453e8cfa11ef3241f  (2018-07-01_13
.17.01)_CIFT_RESULT.zip
Jamess-MacBook-Pro-2:DFRWS jamesma$ ▌
```

3.1.4   Wink Hub

Wink[2] is a brand of software and hardware products that connects and controls smart home devices from a central user interface. Wink connects with third-party smart home devices associated with the Internet of Things. Wink integrates with software from automated home device brands. It supports most smart home devices with zigbee, zwave protocols. This device is the central hub for the smart home devices and is a centralized interface for user accessibility.

3.1.5   Network Pcap

The network file that was captured during evidence collection is a packet capture (pcap) of all the network traffic within the drug lab. The pcap file can be analyzed using Wireshark. The pcap file shows all the network traffic from devices connected to the network at the drug lab.

3.1.6   Samsung Galaxy S6 Edge

Mobile device found at the drug lab. It is unknown who the owner is at the time of acquisition, but the mobile device is connected to all of the IoT devices. Further analysis of the device may reveal crucial data and evidence that was not physically acquired at the crime scene.

---

[2]https://en.wikipedia.org/wiki/Wink_(platform).

3.2 Objective

The objective of the investigation is to analyze the data found to create a timeline of events, determine what time the lab was raided, find out if Jessie Pinkman's two associates, D. Pandana and S. Varga, were involved in the raid, and find out how the QBee Camera was disabled. As part of the objective, we will identify the devices IP address, MAC address, and try and find a serial number. With the information discovered in this process, we hope to paint a digital picture of the devices and their network activity.

4 Forensic Analysis

4.1 Analysis Tools

The below table shows the list of tools that were used for the analysis

Tool	Version	Usage
Autopsy	4.9.1	Data acquisition
Bulk extractor	1.4	Image analysis, file system data carving
Wireshark	2.6.5	Network analysis
Access data FTK imager lite	3.1.1	Data preview and imaging tool
SQLite browser	3.11.0	Database browser

4.2 Network Analysis

The following is a network diagram painting a digital scene. This diagram shows the devices, MAC address, and IP addresses. From the analysis of the PCAP file, we determined all of the devices "spoke" to each other through the raspberry pi ap. We believe this was configured as a router that allowed the various IoT components to communicate back to Jessie's phone in the event that they were set off by either a sensor, or communication through the Echo or Nest Protect. The Wink Hub information is still MIA in regards to the IP and MAC address, and the role it played in the setup of Jessie's lab. With the product information available on the internet, we can determine that the device is used to allow IoT devices to communicate to each other using the same wireless language, making it easier for the consumer to control their various IoT devices. We did however, were able to extract a photo of

SUPERLAB NETWORK DIAGRAM

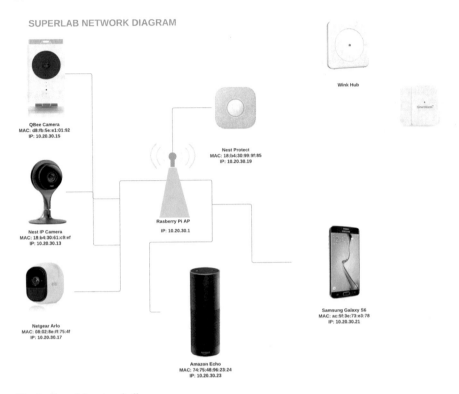

Fig. 1 Superlab network diagram

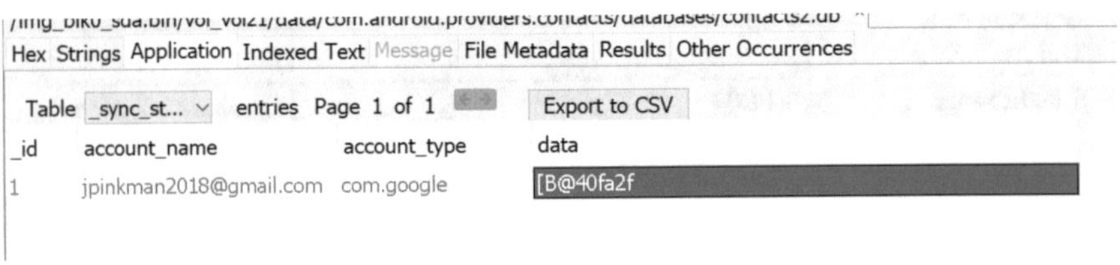

Fig. 2 Email registered to Samsung Galaxy S6 extracted with Autopsy

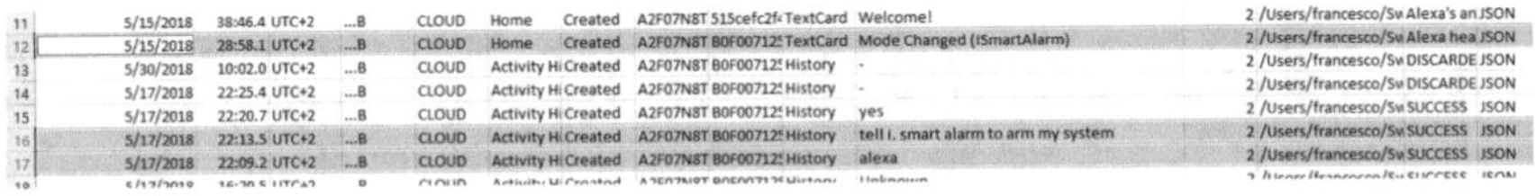

Fig. 3 Alexa voice commands for iSmartAlarm

the device info from Jessie's Samsung Phone, this is exhibited in Fig. 1. In regards to the WinkHub, through the Samsung phone, we found that, Jessie Pinkman, took a photo of his model and serial number, and the Wi-Fi(2.4G) ID number. Those photos are exhibited in Figs. 2, and 3.

⟩ New Database	⟩ Open Database	⟩ Write Changes	⟩ Revert Changes											
Database Structure	Browse Data	Edit Pragmas	Execute SQL										New Record	Delete Record
Table: keys														
andle	key_name	key_form	key_type	account	master_key	public_key	private_key	active_status	creation_time	exportable	expiration_time			
	Filter	Filter	Filter	Filter	Filter	Filter	Filter	Filter	Filter	Filter	Filter			
1	NIX2tIeQ PublicKey	1	P256	Jpinkman2018...		CAESRAogfFBfXDsZ...	MIGHAgEAMB...	1	15263775474...	1	23152995474...			
2	authzen	2	RAW256	Jpinkman2018...	WJQV_x1GVv...			1	15263775482...	2	15289695486...			

Fig. 4 Encryption key found on extracted Samsung Galaxy file

4.3 Forensics Analysis of Gathered Evidence/Devices

4.3.1 Raspberry PI Router

From the findings from the pcap file and by creating a digital picture of the crime scene, we determined that the raspberry pi router was the main device providing network connectivity, and routing for the various IoT devices in the lab. Our analysis of the router focused on painting a digital picture of the traffic that went in and out of the router to create an accurate timeline surrounding the events that occured on the day the drug lab was burned down. Using the IP found in the pcap as 10.20.30.1 with MAC address of B8:27:EB:0E:38:45, we were able to determine direct connections to the Arlo Base Station, Nest Protect, Amazon Echo, and QBEE Camera.

Device	Mac address	IP address
Raspberry PI	b8:28:eb:0e:3b:45	10.20.30.1
QBee Camera	d8:fb:5e:e1:01:92	10.20.30.15
Nest Camera	18:b4:30:61:c9:ef	10.20.30.13
Arlo Pro	08:02:8e:ff:75:4f	10.20.30.17
Nest Protect	18:b4:30:99:9f:85	10.20.30.19
Amazon Echo	74:75:48:96:23:24	10.20.30.23
Samsung Galaxy S6 Edge	ac:5f:3e:73:e3:78	10.20.30.21

4.3.2 Samsung Galaxy S6 Edge

With the recovered Samsung we were able to use forensics software to recover the .bin files. Those files provided the investigators with evidence surrounding the drug lab, movements of Jessie Pinkman, and the IoT devices that were linked to the phone. Specifically, blk0_sda.bin, contained a bulk of information included photos, contacts, the owner's email address, and alerts from the IoT devices such as the Arlo camera, and QBEE camera. This is a screenshot from the Samsung phone showing the owner of the phone's email address. As you can see from (Fig. 2) the email address is, jpinkman2018@gmail.com, from that we can simply deduce that the owner is indeed Jessie Pinkman (Fig. 4).

☑ ⊚ (2018-05-17T10_22_24+0200)_TEXT(TR...
⊚ (2018-05-30T11 09 58+0200) TEXT(TR...

Fig. 5 Alexa Voice file found on application package

Application Packages found on device

Device	Package name	Path
Nest	com.nest.android	/img_blk0_sda.bin/vol_vol21/data/
Netgear	com.netgear.android	/img_blk0_sda.bin/vol_vol21/data/
Wink	com.quirky.android.wink.wink	/img_blk0_sda.bin/vol_vol21/data/
Qbee	com.vestiacom.qbeecamera	/img_blk0_sda.bin/vol_vol21/data/
iSmart	iSA.common	/img_blk0_sda.bin/vol_vol21/data/

Media Devices found on device

File name	Path
687abd91b491c2a...	/img_blk0_sda.bin/vol_vol21/data/com.netgear.android/cache/cams/
af6a076c76e583...	/img_blk0_sda.bin/vol_vol21/data/com.netgear.android/cache/cams/

Amazon Echo

The Amazon Echo, which was placed nearby the door of the lab, is a smart speaker product from Amazon that combines voice recognition "intelligent assistant" capabilities with speaker functionality. From some logs found in the echo, we were able to recover the owner, Jessie Pinkman, talking to the Echo around the time of the raid. As shown below in Fig. 3, Jessie Pinkman is telling Alexa to 'Arm his system" around 10:13 just before the raid happened. There was also voice recorded files of Alexa picking up sounds on what seems to be a fire alarm going off in the background, as shown in Fig. 5.

4.3.3 Wireshark Pcap File

The overall point of that came from the wireshark capture was to be able to see and assign IP and Mac addresses to each of the devices and when they communicate. Below, in Figs. 6 and 7 showing the ARP table of the devices being recognized. You can see that a lot of the communication is between 10.20.30.13(Nest Labs) and 35.195.59.182 (Raspberry Pi. Another thing that stood out was that, in Fig. 8. we see that some ports have been "unreachable", mainly coming from the host. Involving IP address **QBee Camera: 10.20.30.15**, this could be a result of the QBee camera being disabled.

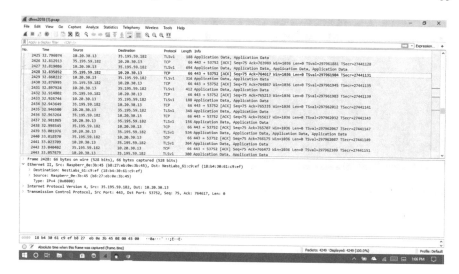

Fig. 6 Network Capture depicting communication between Raspberry Pi router and Nest Camera

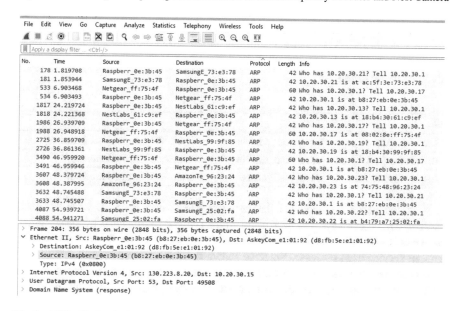

Fig. 7 ARP table for devices connected to the network

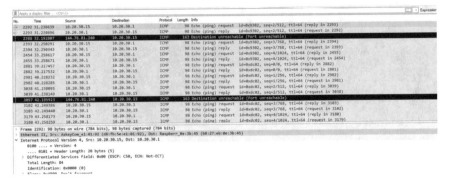

Fig. 8 Network Capture depicting ping requests for qBee Camera

4.3.4 ISmartAlarm

An ISmart device was located at the scene of the crime and was confiscated as evidence. We were given the diagnostics and the dump folders that contained the streams and images for the device. By using autopsy, we were able to find logs of the device's activity for 3 days along with the names of the operators (Pinkman, Vara, and Pandana) as seen in Fig. 9. This provides us with evidence that these people were involved in the crime.

File name	Path
iSmartAlarm.db	/img_blk0_sda.bin/vol_vol21/data/iSA.common/databases
20180517192933771.txt	/img_blk0_sda.bin/vol_vol21/media/0/iSmartAlarm/Log/
20180517093915735.txt	/img_blk0_sda.bin/vol_vol21/media/0/iSmartAlarm/Log/

Next, we were able to locate a database that had Jessie's public key for email and another key. These keys can be helpful in decrypting emails send by Jesse. While we weren't sure what they second key was for, we theorize that it could be a part of another service used by Jesse. We have shown the database in Fig. 4.

Finally, we found some information that clarifies that the door sensor and motion sensor are a part of the iSmartAlarm system. We can prove this through the image provided in Fig. 10. As you can see, it refers to the iSmartAlarm app which proves that these two sensors are a part of the alarm system.

Since these sensors are a part of the ismart system, we can use that information to dig further into the "iso" and search for any logs and activity sent from the sensors to the ismart. As we continue our investigation, we hope to extract more information from the ismart device as we currently have.

Two text files were located within the log folder of the iSmart logs. The content of the text files distinctly display a timestamp in Central European Time.

Table: TB_IPUDairy   ▾ ⓢ ⓐ   New Record   Delete Record

	date	action	IPUID	logType	sensorName	operator	sensorType	sensorID	userID	profileid	profileName
	Filter	Filter	Filter	Filter	Filter	Filter	Filter	Filter	Filter	Filter	Filter
1	2018-05-17 08:37:52		004D3209D9E4	2		pandadodu				2	DISARM
2	2018-05-17 08:34:31		004D3209D9E4	2		pandadodu				2	DISARM
3	2018-05-17 08:34:17		004D3209D9E4	2		TheBoss				1	HOME
4	2018-05-17 08:22:30		004D3209D9E4	2		TheBoss				2	DISARM
5	2018-05-17 08:22:22		004D3209D9E4	2		JPinkman				0	ARM
6	2018-05-17 08:09:57		004D3209D9E4	2		TheBoss				2	DISARM
7	2018-05-17 07:47:50		004D3209D9E4	2		JPinkman				0	ARM
8	2018-05-17 07:45:22		004D3209D9E4	2		TheBoss				2	DISARM
9	2018-05-16 13:55:27		004D3209D9E4	2		JPinkman				0	ARM
10	2018-05-16 13:55:17		004D3209D9E4	2		JPinkman				2	DISARM
11	2018-05-16 13:55:11		004D3209D9E4	2		JPinkman				2	DISARM
12	2018-05-16 13:55:07		004D3209D9E4	2		JPinkman				2	DISARM
13	2018-05-16 13:53:31		004D3209D9E4	2		JPinkman				0	ARM
14	2018-05-16 13:53:10		004D3209D9E4	2		JPinkman				1	HOME
15	2018-05-16 13:47:40		004D3209D9E4	2		TheBoss				2	DISARM
16	2018-05-16 13:47:33		004D3209D9E4	2		TheBoss				2	DISARM
17	2018-05-16 13:47:18		004D3209D9E4	2		TheBoss				1	HOME
18	2018-05-16 13:47:15		004D3209D9E4	2		TheBoss				3	PANIC
19	2018-05-16 13:47:14		004D3209D9E4	2		TheBoss				3	PANIC
20	2018-05-15 13:05:27		004D3209D9E4	2		JPinkman				2	DISARM
21	2018-05-15 13:04:43		004D3209D9E4	2		pandadodu				1	HOME
22	2018-05-15 13:03:47		004D3209D9E4	2		TheBoss				2	DISARM
23	2018-05-15 13:03:43		004D3209D9E4	2		pandadodu				3	PANIC
24	2018-05-15 13:03:36		004D3209D9E4	2		pandadodu				2	DISARM
25	2018-05-15 13:03:28		004D3209D9E4	2		pandadodu				3	PANIC
26	2018-05-15 13:03:11		004D3209D9E4	2		pandadodu				3	PANIC
27	2018-05-15 12:41:35		004D3209D9E4	2		JPinkman				2	DISARM
28	2018-05-15 12:38:20		004D3209D9E4	2		JPinkman				0	ARM
29	2018-05-15 09:38:53	2	004D3209D9E4	5							
30	2018-05-15 09:34:20	1	004D3209D9E4	5							

Fig. 9 iSmart database showing who accessed the device

```
TheMotion - Motion Detector
Time
11:08 AM
Date
April 25, 2018
Phone Numbers Notified
0792245315
Please log into your iSmartAlarm App immediately to review activity.
Questions? Need troubleshooting help?
Please visit our Support Center
SUPPORT CENTER
keywords"
```

```
TheBouncer - Door Sensor
Time
11:06 AM
Date
April 25, 2018
Phone Numbers Notified
0792245315
Please log into your iSmartAlarm App immediately to review activity.
Questions? Need troubleshooting help?
Please visit our Support Center
SUPPORT CENTER
keywords"
```

Fig. 10 Encryption key found on extracted Samsung Galaxy file

4.3.5 WinkHub

A winkhub device was found at the scene of the crime, thus making it evidence. The main purpose of a winkhub is to control other smart devices. Since the main way to use it through the phone, a majority of the connection is between the samsung phone and the Winkhub device.

To start, we were able to find this.png file in the zigbee-ota-images folder and was modified May 19 as shown in Fig. 11. We still require further research to understand this picture and whether this will assist in the crime examination.

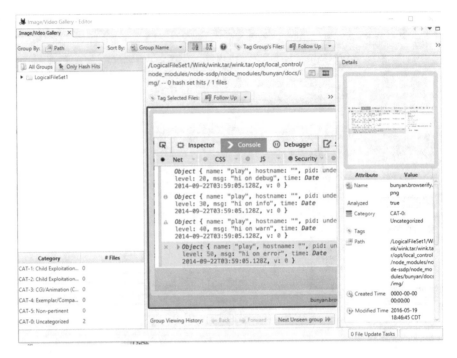

Fig. 11 JSON file found for Wink

We were also able to find email addresses, roughly 95 in total. A majority just give us scripts, while some provide us with.json files. We also located a few me@ email addresses which could be from the admin of the Winkhub. In Fig. 12, we display two lists from Autopsy.

4.3.6 Arlo

This is a snapshot from the phone that showed an alert being pinged to the phone from the Arlo camera. The alert came on the 17th of May 2018 between the times of 17:32:50 UTC—17:33:22 UTC. The camera showed multiple individuals moving freely throughout the lab (Fig. 13).

Several cache images were found within the Netgear APK. These images were system generated, and images used for preview purposes while using the Arlo application interface. The images show a sequence of events on the date of May 17, 2018. Two time image time frames were acquired. 07:39:33 UTC & 17:32:20 UTC. During these time frames shown multiple individuals were moving freely around the lab.

Fig. 12 Email list for Wink

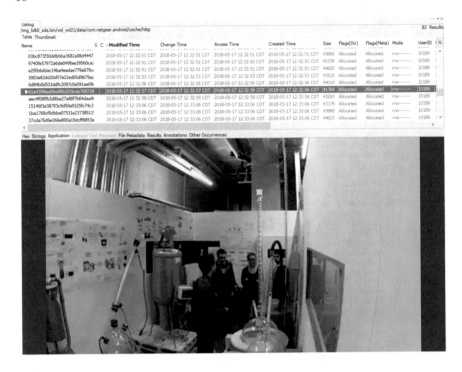

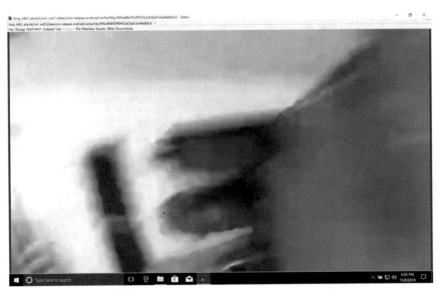

Fig. 13 Cached images captured by Arlo camera

This Image shows an unknown individual that is messing with something in the corner of the synthesis room. According to the drug lab diagram. The Qbee camera is positioned in that area.

4.4 Images Found on Samsung Galaxy S6 Edge

See Figs. 14 and 15.

Fig. 14 Image of WinkHub
MAC address found on dcim
folder of Samsung Galaxy S6

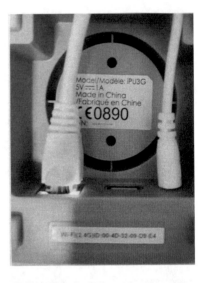

Fig. 15 Image of WinkHub
Model Number found on
dcim folder of Samsung
Galaxy S6

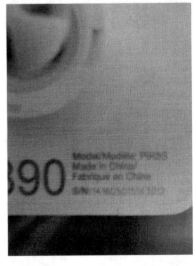

DFRWS IoT Forensic Challenge Report 4

Mikaela Zabel, Alex Molina and Monica de Leon Gamon

Abstract In this chapter, we report on the findings of our group's analysis of the Digital Forensic Research Workshop (DFRWS) IoT Forensic Challenge (2018–2019).

1 Report Summary

An illegal drug lab was set on fire. Police and digital forensic specialist arrive on scene but the owner, Jessie Pinkman, was missing.

1.1 Challenge Questions from Attorney

1.1.1 At What Time Was the Illegal Drug Lab Raided?

On May 17, 2018, the police were called at 10:40 am, and the police and digital forensic specialist arrived by 10:45 am. We know that the raid took place earlier that day before 10:40 am. The audio file at 10:22 am had a beeping sound, possibly indicating a fire alarm going off. It is possible the fire alarm was going off at 10:22, indicating the raid was right before then.

M. Zabel · A. Molina · M. de Leon Gamon (✉)
Department of Information System and Cyber Security,
University of Texas at San Antonio, San Antonio, TX, USA
e-mail: monicadlg96@gmail.com

M. Zabel
e-mail: khc347@my.utsa.edu

A. Molina
e-mail: mmolina2296@gmail.com

© Springer Nature Switzerland AG 2020
X. Zhang and K.-K. R. Choo (eds.), *Digital Forensic Education*,
Studies in Big Data 61, https://doi.org/10.1007/978-3-030-23547-5_5

Table 1 Deductions made from collected evidence

Deductions
(1) According to Pandana and Varga's interrogation, both declined any involvement in the crime scene. However, according to the pictures (Fig. 15) found in the Samsung device, there were three different men operating in the lab. Based on this evidence, we can confirm that three men might be involved in the raid
(2) We could not confirm the time and date when the pictures were taken
(3) According to the Amazon Echo's file (Fig. 7), "Avenue Forel, Lausanne, VD, CH" could be the physical address of the site or an address of a suspect involved in the crime scene
(4) On the day of the incident, there is an audio file (Fig. 10) found on the Amazon Echo, indicating a beeping sound which is presumed to be the smoke alarm. Within 18 min of the smoke alarm going off, perhaps someone from the outside noticed the building catching fire and called the police on the illegal lab
(5) Based on the iSmartAlarm database (Fig. 22), we found that the user "pandadodu," disabled the iSmartAlarm at 5:34 am. For this reason, we assume that Pandana was involved in the raid

1.1.2 Could Any of Jessie Pinkman's Friends Have Been Involved in the Raid?

Below is a table of deductions we made based on the collected evidence from the crime scene.

Due to a username of 'pandadodu' disabling the ismartalarm device at 5:34 am and 10:37 am the day of the attack, so it is possible that Jessie Pinkman's friend D. Pandana was involved in the raid (Table 1).

1.1.3 What Is the Confidence in Such Hypothesis?

The confidence in this hypothesis that Jessie Pinkman's friend D. Pandana is plausible. While there are records found on the samsung device (Fig. 22) that a username of 'pandadodu' was used to disarm the iSmartAlarm device at 5:34 am and 10:37 am the day of the raid, there would need to be confirmation from the manufacturing company of the ownership of the account.

1.1.4 How Was the QBee Camera Disabled?

Information providing insight about the QBee camera was not uncovered within the devices listed above. While this is unfortunate, a possible location of future investigation to how the QBee camera was disabled may be found within the samsung device at/img_blk0_sda.bin/vol_vol21/data/com.vestiacom.qbeecamera/databases. Current efforts in investigations were hindered with the databases appearing to be encrypted.

2 Details of Scenario[1]

- 10:40 2018-05-17

  - The police received an alert that an illegal drug lab was set on fire.

- 10:45 2018-05-17

  - Police and a digital forensic specialist arrive on scene. Jessie Pinkman, the owner of the illegal drug lab, is missing.

- Police interrogate two of Jessie Pinkman's associates: D. Pandana and S. Varga. Pandana and Varga admit having access to the drug lab's WiFi network but decline any involvement in the raid. They say that Jessie Pinkman had the IoT security systems installed because he feared attacks from a rival gang and that Jessie kept the alarm engaged in "Home" mode whenever he was inside the drug lab.

  Digital forensic specialist observes some digital devices:

  1. Alarm system (iSmartAlarm)
  2. Smoke detector (Nest Protect)
  3. Amazon Echo
  4. Raspberry Pi
  5. Samsung Mobile Device
  6. Samsung Device
  7. QBee Camera
  8. Netgear Device
  9. Wink Hub
  10. Three cameras (QBee Camera, Nest Camera, and Arlo Pro.

3 Timeline of Events

See Table 2.

4 Concept Diagram

Figure 1 is a picture representing the Internet of Things architecture with all the digital devices used, according to the digital forensic specialist. The Samsung device provided visual evidence with pictures from within the drug lab along with photos of devices including MAC addresses and IP addresses. The Amazon Echo provided time stamps of audio clips from inside the drug lab. It also provided valid email addresses

---

[1] https://www.dfrws.org/dfrws-forensic-challenge.

Table 2 Timeline of events

Date	Time	Activity	Device(s)
03 Mar 2018	05:12	Personal email accessed (jpinkman2018@gmail.com)	Samsung device
26 Mar 2018	16:49	Photo of netgear arlo base station	Samsung device, arlo camera
10 Apr 2018	09:19–20	Three photos of NestLabs device (Figs. 17 and 19)	Samsung device, NestLabs device
10 Apr 2018	09:21	Photo of nest camera device (Fig. 20)	Samsung device, nest camera
02 May 2018	13:29	Photo of QBee camera (Fig. 21)	Samsung device, QBee camera
15 May 2018	11:28	iSmartAlarm status changed on alarm mode	iSmartAlarm, Amazon Echo
17 May 2018	10:22	iSmartAlarm status changed on alarm mode (Fig. 10)	iSmartAlarm, Amazon Echo
17 May 2018	10:22	Sound byte of smoke alarm on	Amazon Echo, Samsung device
17 May 2018	10:40	Police alerted of drug lab invaded and set on fire	
17 May 2018	10:45	Police and digital forensic team arrive at crime scene	
17 May 2018	18:44	Photo of Wink Hub device (Fig. 23)	Samsung device, Wink Hub
17 May 2018	19:32	Visual activity captured on camera of men in lab (Fig. 14)	Samsung device, Arlo camera

and possible physical addresses associated with the account and the potential suspects involved in the drug lab. The iSmartAlarm provided our team to better visualize the ongoings inside the lab and create possible scenarios of how and when the lab was raided on May 17, 2018.

5 Analysis Results

See Table 3.

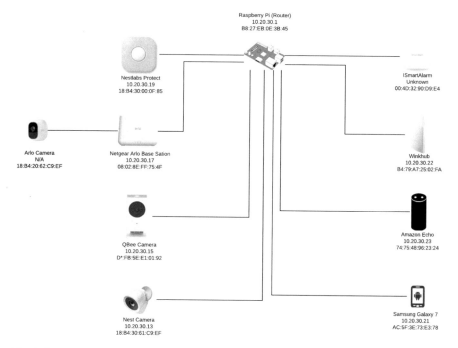

Fig. 1 Internet of things (IoT) forensics

Table 3 Digital devices

Name	IP	Mac address
Netgear arlo base station	10.20.30.17	08:02:8e:ff:75:4f
Netgear arlo camera	N/A	18:B4:20:61:C9:EF
NestLabs camera	10.20.30.13	18:b4:30:61:c9:ef
NestLabs protect	10.20.30.19	18:b4:30:99:9f:85
Amazon Echo	10.20.30.23	74:75:48:96:23:24
Samsung (S7)	10.20.30.21	ac:5f:3e:73:e3:78
Samsung (Wink Hub)	10.20.30.22	b4:79:a7:25:02:fa
Raspberry Pi	10.20.30.1	b8:27:eb:0e:3b:45
QBee camera	10.20.30.15	d8:fb:5e:e1:01:92
Motion sensor	N/A	N/A
Contact sensor	N/A	N/A

5.1 Network Captured Data

Raspberry Pi 10.20.30.1

178 1.819708	Raspberr_0e:3b:45	SamsungE_73:e3:78	ARP	42 Who has 10.20.30.21? Tell 10.20.30.1
181 1.853944	SamsungE_73:e3:78	Raspberr_0e:3b:45	ARP	42 10.20.30.21 is at ac:5f:3e:73:e3:78

Samsung (Wink Hub) 10.20.30.22

4087 54.939721	Raspberr_0e:3b:45	SamsungE_25:02:fa	ARP	42 Who has 10.20.30.22? Tell 10.20.30.1
4088 54.941271	SamsungE_25:02:fa	Raspberr_0e:3b:45	ARP	42 10.20.30.22 is at b4:79:a7:25:02:fa

Samsung (S7) 10.20.30.21

178 1.819708	Raspberr_0e:3b:45	SamsungE_73:e3:78	ARP	42 Who has 10.20.30.21? Tell 10.20.30.1
181 1.853944	SamsungE_73:e3:78	Raspberr_0e:3b:45	ARP	42 10.20.30.21 is at ac:5f:3e:73:e3:78

Netgear Arlo Base Station 10.20.30.17

1986 26.939709	Raspberr_0e:3b:45	Netgear_ff:75:4f	ARP	42 Who has 10.20.30.17? Tell 10.20.30.1
1988 26.948918	Netgear_ff:75:4f	Raspberr_0e:3b:45	ARP	60 10.20.30.17 is at 08:02:8e:ff:75:4f

NestLabs Camera 10.20.30.13

1817 24.219724	Raspberr_0e:3b:45	NestLabs_61:c9:ef	ARP	42 Who has 10.20.30.13? Tell 10.20.30.1
1818 24.221368	NestLabs_61:c9:ef	Raspberr_0e:3b:45	ARP	42 10.20.30.13 is at 18:b4:30:61:c9:ef

NestLabs Protect 10.20.30.19

2725 36.859709	Raspberr_0e:3b:45	NestLabs_99:9f:85	ARP	42 Who has 10.20.30.19? Tell 10.20.30.1
2726 36.861361	NestLabs_99:9f:85	Raspberr_0e:3b:45	ARP	42 10.20.30.19 is at 18:b4:30:99:9f:85

Amazon Echo 10.20.30.23

3607 48.379724	Raspberr_0e:3b:45	AmazonTe_96:23:24	ARP	42 Who has 10.20.30.23? Tell 10.20.30.1
3608 48.387995	AmazonTe_96:23:24	Raspberr_0e:3b:45	ARP	42 10.20.30.23 is at 74:75:48:96:23:24

QBee Camera 10.20.0.15

4203 56.278903	AskeyCom_e1:01:92	Raspberr_0e:3b:45	ARP	42 Who has 10.20.30.1? Tell 10.20.30.15
4204 56.278923	Raspberr_0e:3b:45	AskeyCom_e1:01:92	ARP	42 10.20.30.1 is at b8:27:eb:0e:3b:45

5.2 Arlo Camera

Figure 2 was found in the *arlo_nand.tar.gz* system log's file. This image provides the Arlo Camera IP Address: 10.20.30.17.

Figure 3 was found in the *arlo_nand.tar.gz* xlog's file, which provides the Arlo Camera MAC address: 08:02:8E:FD:BD:CD.

Fig. 2 SysLog file

Fig. 3 Xlog file

5.3 Amazon Echo

Figures 4, 5 and 6, represent a view into the *cift_amazon_alexa* file using Autopsy to exfiltrate the data. In the second image (Fig. 5), the Indexed Text tab outlines verbal cues made by the suspects into the Echo device. Figure 6 outlines the personal email linked with the Amazon Echo account associated with the suspect. The email listed is jpinkman2018@gmail.com.

The following image was found in the file *cift_amazon_alexa_ALEXA_ DEVICE.csv*. The image below (Fig. 7) has a Swiss address highlighted, relating to the "locale" variable outlined above it. "Avenue Forel, Lausanne, VD, CH" is a real physical address in Zurich, Switzerland. It appears to be linked to Jessie's Alexa Apps folder.

The following information was found in the *cift_amazon_alexa_SETTINGS_ MISC.csv*. Figure 8 identifies a physical address associated with the possible third party accounts iHeartRadio, Tune In, and Amber. According to the Amazon Echo

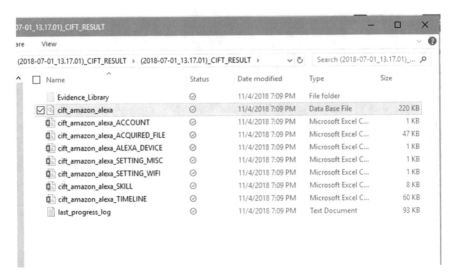

Fig. 4 *cift_amazon_alexa* file

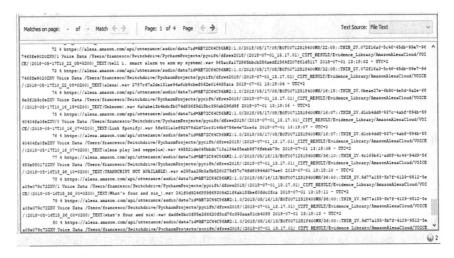

Fig. 5 Echo device data

Settings' file, "4204 Colby Ave., Everett WA 98203" could be the physical address of the site or an address of a suspect involved in.

Figures 9, 10, 11 and 12 were found in the file of *cift_amazon_alexa_SKILL.csv*. The following images outline the linked accounts associated with the Amazon Echo device. It has the Arlo camera, the Nest Camera, the Wink device, and the iSmartAlarm device linked.

The below image (Fig. 13) is a screenshot of the audio voice files within the Amazon Echo device. In each file name is the text audio translated into text. A

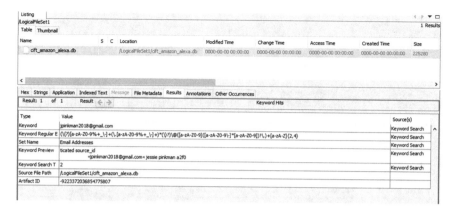

Fig. 6 Personal email linked to Echo device

Fig. 7 *cift_amazon_alexa_ALEXA_DEVICE.csv* file

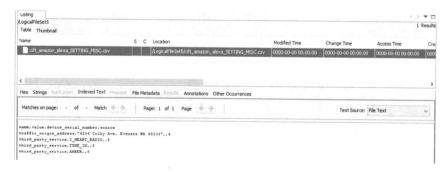

Fig. 8 Physical address associated with third party accounts

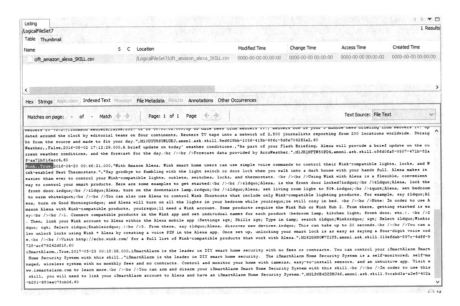

Fig. 9 Amazon Echo device linked to arlo devices

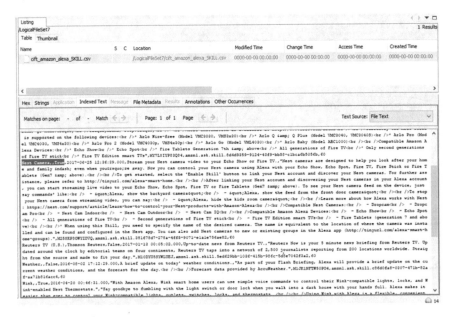

Fig. 10 Amazon Echo device linked nest devices

Fig. 11 Amazon Echo device linked to wink devices

Fig. 12 Amazon Echo device linked to iSmartAlarm devices

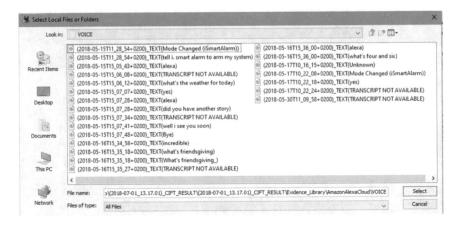

Fig. 13 Audio files

relevant audio file appears to be the very first audio clip indicating a change in status of the security alarm. This audio clip occurred on 05-17-2018 at 10:22:20 according to the file.

Another relevant audio file would be the second to last file indicating "TRAN-SCRIPT NOT AVAILABLE" at 10:22:24 on 05-17-2018. When opened and listened to, the audio indicates a beeping which can be inferred to be the smoke alarm. This occurred 18 min before the police were called about a fire at the laboratory. With this information, we assume that the drug lab was raided at this time and approximately 18 min had gone by until a neighbor or passer-byer noticed the fire and alerted the authorities.

The other audio files are not relevant to the crime. The files are found in *cift_amazon_alexa_TIMELINE.csv*.

5.4 Samsung Device (S6)

The information in Figs. 12, 13, 14, and 15 is helpful because the email serves as a target selector and can be directly related to the crimes. This email address will assist in retrieving more information on the suspect and the on-goings at the site.

Figure 16 were found in the files on the Samsung device. These snapshots provide direct evidence of the inside of the crime scene and provide images of the suspects involved. According to these images of the drug lab, we assume that three different men were operating in the lab. We do not have an ID match on these suspects, but it would be up to the police to match them to Pandana and Vargas to confirm their involvement.

Images found in Samsung files:

▼ 🗀 img_blk0_sda.bin

    ▼ 🗀 vol_vol21

        ▼ 🗀 data

            ▼ 🗀 com.netgear.android

                🗎 **files (1)**

                🗎 shared_prefs (1)

            ▼ 🗀 com.google.android.gm

                🗎 **app_sslcache (1)**

            ▼ 🗀 cache

                🗎 **jpinkman2018@gmail.com (2)**

Fig. 14 Pinkman's personal email

Type	Value	Source...
URL	https://inbox.google.com/accounts/SetOSID?authuser=0&continue=https%3A%2F%2Finbox.google.com%2F%3Fpli%3D1&osidt=ALWU2cuzXI8vadmxJJkzpn2ArZ6rTOMW63fT9spEisIFAuNHFBl_nGyns5Uv1SZTleaGGJY6HYgfc9VMYN67A5-ms9-eK4xS_KoFhwNzmtMWYjOKFkmebLFs3Ari8tCm7Kbk2adYtJ8u	Recent Activity
Date Acce	2018-03-16 05:12:06	Recent Ac
Referrer URL	https://inbox.google.com/accounts/SetOSID?authuser=0&continue=https%3A%2F%2Finbox.google.com%2F%3Fpli%3D1&osidt=ALWU2cuzXI8vadmxJJkzpn2ArZ6rTOMW63fT9spEisIFAuNHFBl_nGyns5Uv1SZTleaGGJY6HYgfc9VMYN67A5-ms9-eK4xS_KoFhwNzmtMWYjOKFkmebLFs3Ari8tCm7Kbk2adYtJ8u	Recent Activity
Title	Inbox – jpinkman2018@gmail.com	Recent Ac
Program N	Chrome	Recent Ac
Domain	inbox.google.com	Recent Ac
Source File Path	/img_blk0_sda.bin/vol_vol21/data/com.android.chrome/app_chrome/Default/History	
Artifact ID	-9223372036854775807	

Fig. 15 Activity associated with Jessie Pinkman's email

Fig. 16 Arlo snapshots captured from the Samsung smart phone

Fig. 17 Image of NestLabs device

5.4.1 Photos from Samsung Device

Figure 17 is an image of the Netgear Base Station.

Figure 18 is an image of the NestLabs smoke detector found on the Samsung device. This is the device that the Amazon Echo captured when the alarm went off.

Figure 19 is an image of the NestLabs Hub device found on the Samsung device.

Figure 20 is an image of an unknown device found on the Samsung device.

Figure 21 is an image of a Nest Camera found on the Samsung Device.

Figure 22 is a snapshot image found on the Samsung Device.

The above screenshot outlines usernames of the operators arming and disarming the iSmartAlarm device. The three usernames used are "pandadodu", "TheBoss", and "JPinkman". Jessie Pinkman can very likely be deduced to be using "JPinkman". A

2018-04-10 02:19:25 CDT | 2018-04-10 02:19:25 CDT | 2018-04-10 02:19:24 CDT | 2018-04-10 02:19:24 CDT

Fig. 18 Image of nest camera

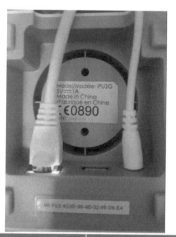

2018-04-10 02:20:15 CDT | 2018-04-10 02:20:15 CDT | 2018-04-10 02:20:15 CDT | 2018-04-10 02:20:15 CDT

Fig. 19 Image of NestLabs device plugs

possible associate, D. Pandana, is closely related to the username "pandadodu". We can possibly assume Jessie Pinkman was using a second account "TheBoss", but we cannot verify that it belongs to him (Fig. 23).

2018-04-10 02:20:59 CDT | 2018-04-10 02:20:59 CDT | 2018-04-10 02:20:59 CDT | 2018-04-10 02:20:59 CDT

Fig. 20 Image of NestLab device model

2018-04-10 02:21:20 CDT | 2018-04-10 02:21:20 CDT | 2018-04-10 02:21:20 CDT | 2018-04-10 02:21:20 CDT

Fig. 21 Image of nest camera

5.5 Alarm System (ISmart)

The data tables above outline a time discrepancy found in the time stamps and the actual local time. The time stamps shown are seven (7) hours behind the local time. In the timeline table on page 2, the time stamps are already modified to account for this time difference (Fig. 24).

Fig. 22 Image of QBee camera technical details

	rowid	date	operator	profileid	profileName
1	1	1526546272	pandadodu	2	DISARM
2	2	1526546071	pandadodu	2	DISARM
3	3	1526546057	TheBoss	1	HOME
4	4	1526545350	TheBoss	2	DISARM
5	5	1526545342	JPinkman	0	ARM
6	6	1526544597	TheBoss	2	DISARM
7	7	1526543270	JPinkman	0	ARM
8	8	1526543122	TheBoss	2	DISARM
9	9	1526478927	JPinkman	0	ARM
10	10	1526478917	JPinkman	2	DISARM

Fig. 23 Database depicting ISmartAlarm device disarmed by username 'pandadodu' at Thursday, May 17, 2018 5:34:31 AM and Thursday, May 17, 2018 10:37:52 AM

✗ 1525269028098.jpg /img_blk0_sda.bin/vol_vol21/media/0/DCIM/.thumbnails/1525269028098.jpg

```
Title:20180517192933772

2018-05-17  19:29:33.773    1526578173773
UserName:iSmartAlarm  2.0.8

TimeZone:Central European Standard Time

Time:2018-05-17  19:29:33.773

phone model:SM-G925F
```

△ Modified Time	Change Time	Access Time	Created Time
2018-05-17 12:36:04 CDT	2018-05-17 12:36:04 CDT	2018-05-17 12:36:04 CDT	2018-05-17 12:36:04 CDT
2018-05-17 12:36:04 CDT	2018-05-17 12:36:04 CDT	2018-05-17 08:36:14 CDT	2018-05-17 08:36:14 CDT
2018-05-17 12:38:24 CDT	2018-05-17 12:38:24 CDT	2018-05-17 12:38:24 CDT	2018-05-17 12:38:24 CDT
2018-05-17 12:38:32 CDT	2018-05-17 12:38:32 CDT	2018-05-17 12:29:33 CDT	2018-05-17 12:29:33 CDT

Fig. 24 iSmartAlarm data found on the samsung device

5.6 Samsung (Wink Hub)

A private key was found on the Wink Hub device. This information would lead our team to believe that asymmetric handshakes were taking place. The interactions involved are expected to have a corresponding public key (Figs. 25, 26 and 27).

Fig. 25 Image of Wink Hub device

```
May 17 11:44:30 flex-dvt user.warning hub[561]: WARNING: (readLocalControlThread.c:38) Read Failure - error -2
May 17 11:44:43 flex-dvt user.notice hub[561]: NOTICE: (hubRemoteSocket.c:592) #~#~#1 Remote Cassette ping, topic 1367827
May 17 11:44:43 flex-dvt user.notice hub[561]: NOTICE: (writeThread.c:16) Writing Cassette to Socket: receipt Size: 28, t
May 17 11:44:55 flex-dvt daemon.info udhcpc[1215]: Sending renew...
May 17 11:44:55 flex-dvt daemon.info udhcpc[1215]: Lease of 10.20.30.22 obtained, lease time 600
```

Fig. 26 IP address

Wink Private Key
-----BEGIN PRIVATE KEY-----
MIICdwIBADANBgkqhkiG9w0BAQEFAASCAmEwggJdAgEAAoGBAL9R0WTGOjtSsJhU
dyy3woSNB8bnQF6Z919IIO6tQcb339p+z7x3Wa0UR1nmHNGyquNmKpIVJsqSW0SU
zlazUqOdG995L6sunji0oGCJZfDcYJQgfiTgOAmoX1YIAZ77QnctIJI98H9BNXpz
MKzEZ60omWGERt6mEzambKhnqx2xAgMBAAECgYEAtmI9B6ChKqRtjQoYghy75ri7
TTZu+sA0PjT4kA/CwPbUA+sYBkaQypqmWmEv4Ag/OIDQ41N1o29aakqfn1mb0I+N
IhRXnp3r3Jb4vktYv+7s2ZSq+GeX30qSfecRG1VrQFgjFzFaV/egDNBtX13HPnKH
N3XswQ4jtz1gi1qMy0ECQQDiE7mHk6JlyRzTiRU67873O3JBsANN3nksf0ZVW+ad
TPgTO59s5csLOo5wwZvbHYJpac0JrzJfL7lj8fc3DpJ9AkEA2KRj5J8WbgrCq/K9
Fmka5dSKBkEe/ObrOhf7xpqdNt+5E/D3fdFhtd+hWYoPzNu2PMLv3dGEu05EM1oW
v3PKRQJBAJkFbtOkivinH5rSs4sD3FudYhWyFFp1liEXOLz4CI04RsWF2r/zvf4o
sVcCxQKHs8iz8zN14Stqexp6IbGjbGUCQDjoHbFNNLgdLXu7XH4Jh949UTBJZNIr
0vhEmodVz9w9MkGRt4EeQftjTd/WS0Y1pI4baqSqx6Hn8jRHAtixkr0CQFvtT/z5
sWCnSv0whgNqJyFDZq3EGF8Zs7wqEtKN9NhgO5FlgPWpK1+sZV11afDDEFUGf6xs
scbli+RzURoKQEk=
-----END PRIVATE KEY-----

Wink Certificate
-----BEGIN CERTIFICATE-----
MIIB6jCCAVOgAwIBAgIJAJT8tB5KTfziMA0GCSqGSIb3DQEBBQUAMA4xDDAKBgNV
BAMMA2FhdTAeFw0xNjA1MTUxODQ2NDNaFw0yNjA1MTMxODQ2NDNaMA4xDDAKBgNV
BAMMA2FhdTCBnzANBgkqhkiG9w0BAQEFAAOBjQAwgYkCgYEAv1HRZMY6O1KwmFR3
LLfChI0HxudAXpn3X0gg7q1Bxvff2n7PvHdZrRRHWeYc0bKq42YqmVUmypJbRJTM
hrNSo50b33kvqy6eOLSgYIII8NxgICB+JOA4CahfViUBnvtCdy2UmX3wf0E1enMw
rMRnrSiZYYRG3qYTNqZsqGerHbECAwEAAaNQME4wHQYDVR0OBBYEFAbARwJO0uV2
I820wyrO6SQMveTfMB8GA1UdIwQYMBaAFAbARwJO0uV2I820wyrO6SQMveTfMAwG
A1UdEwQFMAMBAf8wDQYJKoZIhvcNAQEFBQADgYEAMEqW+FCfAjhrybCnsaH9soaC
X5fot61dOYG0Y3Bg1iOI8MCNruT72AVRITr3S654Py6XIOwIlJJkZDq64AODRzui
nXY8Jp8E1hn0hLoOtmwSvGioIkvxhBzDKIc5bOv9ZbYNnY8SDvKgO7YjvtSfBIL1
q6xWTM/9LwHz3fU/3yE=
-----END CERTIFICATE-----

Fig. 27 Private key and certificate from wink device

DFRWS IoT Forensic Challenge Report 5

Shawn Knox, John Chismar and Guo Mah

Abstract In this chapter, we report on the findings of our group's analysis of the Digital Forensic Research Workshop (DFRWS) IoT Forensic Challenge (2018–2019).

1 Introduction

1.1 Case Scenario

An illegal drug lab was invaded on May 17th, 2018. The Attorney General wants to know what time the illegal drug lab was raided and whether or not the owner's (Jesse Pinkman) friends were involved in the raid.

1.1.1 Scenario Details

At 10:40 (UTC +2) police were alerted of the raid. Jesse Pinkman's two associates are interrogated. They admit to having access to the lab's WiFi network but say they were not involved in the raid. The following IoT devices were captured on scene: an iSmartAlarm alarm system, a Qbee Camera, a Nest Camera, an Arlo Pro Camera, and a Nest Protect smoke detector.

1.2 Answering the Attonery's Questions

Currently, with the evidence we have, we believe that the raid occurred from 10:34:31 AM to 10:36:06 AM.

S. Knox (✉) · J. Chismar · G. Mah
Department of Information System and Cyber Security,
University of Texas at San Antonio, San Antonio, TX, USA
e-mail: ShawnPknox11@gmail.com

© Springer Nature Switzerland AG 2020
X. Zhang and K.-K. R. Choo (eds.), *Digital Forensic Education*,
Studies in Big Data 61, https://doi.org/10.1007/978-3-030-23547-5_6

Table 1 Timeline

10:03:43 AM	Nest camera disabled
10:16:58 AM	Nest camera reconnected
10:22:22 AM	Jesse Pinkman arms the iSmartAlarm
10:34:28 AM	Nest camera is alerted that someone is in the lab
10:34:31 AM	iSmartAlarm operator 'Pandadodu' disarms the system
10:35:03 AM	End of nest camera alert
10:36:06 AM	Smoke warning
10:40:00 AM	Police are alerted of a raid
10:45:00 AM	Police are on scene

We also think that out of the two associates, D. Pandana may have been involved in the raid.

At this time, we cannot say how or when the Qbee Camera was disabled.

1.3 Timeline of Events

See Table 1.

2 Network Traffic Analysis

Device 11	IP address	MAC address
Nest camera	10.20.30.13	18:B4:30:61:C9:EF
Nest protect	10.20.30.19	18:B4:30:99:9F:85
Raspberry Pi	35.195.59.182 10.20.30.1 172.217.16.142 172.217.23.106	B8:27:EB:0E:3B:45
QBee camera	10.20.30.15	D8:FB:5E:E1:01:92
SamsungE 73	10.20.30.21	AC:5F:3E:73:E3:78
SamsungE 25	10.20.30.22	B4:79:A7:25:02:FA
Arlo Pro	10.20.30.17	08:02:8E:FF:75:4F
Amazon Echo	10.20.30.23	74:75:48:96:23:24
IPv4mCast	239.255.255.250	01:00:5E:7F:FF:FA

These images show the IP and MAC Address of the devices captured and/or found on the network. We believe this is how traffic moves across the network (Fig. 1).

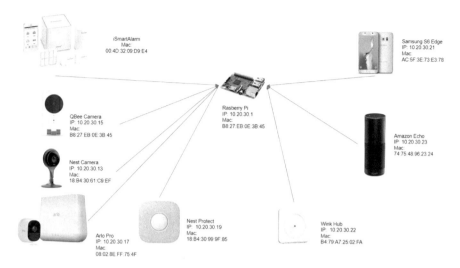

Fig. 1 How traffic moves across the network

3 Device Overview

3.1 Nest Camera

The Nest Camera is an indoor security camera that monitors any activity that happens within the home.

3.1.1 Device Features

- Cloud recording: Doesn't use memory cards to store your videos on the camera. Upload continuously to the cloud. 24/7 cloud recording
- Alerts on your phone: Says when it sees a person and is able to identify familiar faces.

3.1.2 Forensic Analysis

We found that Nest camera actually captured someone and had a snapshot of the person at timestamp 1526546068050, which converts to Thursday May 17, 2018 10:34:28 (am) in GMT +2:00. The end time of the event is at timestamp 1526390072912, which converts to Thursday May 17, 2018 10:35:03 (am) (Fig. 2).

```
"last_cuepoint": {
    "start_time": 1526567213971,
    "image_url_prefix": "https:\/\/www.dropcam.com\/api\/wwn.get_image",
    "image_url_suffix": "cuepoint_id=1526567213;",
    "end_time": 1526567220435,
    "animated_image_url_suffix": "cuepoint_id=1526567213;",
    "id": "1526567213",
    "activity_zone_ids": "",
    "animated_image_url_prefix": "https:\/\/www.dropcam.com\/api\/wwn.get_anima
    "types": "motion"
},
"ip_address": "10.20.30.13",
"activity_zones": [

],
"camera_type": 8,
"streaming_state": "streaming-enabled",
"nexus_api_http_server_url": "https:\/\/nexusapi-eu1.dropcam.com",
"where_id": "00000000-0000-0000-3355-718c34bd39f2",
"serial_number": "18b43061c9ef",
"preview_streaming_enabled": true,
"snapshot_url": {
    "snapshot_url_prefix": "https:\/\/www.dropcam.com\/api\/wwn.get_snapshot",
    "snapshot_url_suffix": ""
},
"public_share_enabled": false,
"person_event": {
    "start_time": 1526546068050,
    "end_time": 1526546103079
},
"streaming_enabled_by_wwn": {
    "state": true,
    "last_changed": 1526390072912
}
}
}
},
{
"object_key": "topaz.18B430000039E345",
"object_revision": -20195,
"object_timestamp": 1526577649180,
"value": {
    "spoken_where_id": "00000000-0000-0000-0000-00010000000a",
    "creation_time": 1526289627039,
    "installed_locale": "en_GB",
    "ntp_green_led_brightness": 2,
    "component_buzzer_test_passed": true,
    "wifi_ip_address": "10.20.30.19",
    "wired_led_enable": true,
```

Fig. 2 Nest camera motion detected /img_blk0_sda.bin/vol_vol21/app/com.nest.android/cache/cache

3.2 Amazon Echo

The Amazon Echo is a hands-free, speaker and smart assistant. It has 7 built in microphones to be able to clearly hear you from across a room. Voice commands allow you to use the device in conjunction with other smart devices around the house. The Echo also connects to other services such as weather tracking websites to give updates over certain locations.

3.2.1 Device Features

- Voice control: Alexa voice-assistant listens for commands and executes actions accordingly.
- Smart Device Connectivity: Is able to connect to other devices on the network such as your phone.

3.3   Arlo Pro

The Arlo Pro is a smart security camera with the ability to capture HD video and audio. Information captured by the device is stored in the cloud or locally backed up. It's also able to connect to smart assistants such as Google Assistant and Amazon Alexa.

3.3.1   Device Features

- Night Vision
- Advanced Motion Detection
- Video streaming
- 2-way audio.

3.4   Nest Protect

The Smart Alarm is a smart smoke alarm. It has the ability to detect fires before they start and what room they started in. The Nest Protect also has path lighting; if someone stands underneath the device it will shine a light for them. You can also connect your phone to the device to silence alerts. Finally, Nest sends monthly reports via email to let you know device activity and other bits of information.

3.4.1   Device Features

- Path lighting
- Phone connectivity
- Location and time based alerts
- Early fire detection.

3.4.2   Forensic Analysis

We also confirmed one more MAC address which we were not able to locate in our last milestone. We found this image in Autopsy, and scanned the QR code. We got a bunch of characters, but the 4th line makes us believe that it is the MAC address for this device. The devices MAC was: 18:B4:30:99:9F:85 (Figs. 3 and 4).

We found that there is a thread called "protect smoke warn", where we believe that it is the warning from the Nest Protect. If we recall the scenario, police were alerted that the lab was invaded and unsuccessfully set on fire at 10:40am. Based on what we found, there is a smoke warning at time stamp 1526546166 (Thursday May 17, 2018 10:36:06 (am)), and the warning got cleared immediately after 14 seconds,

Fig. 3 Nest protect
/img_blk0_sda_bin/vol_vol21/
media/0/DCIM/Camera/
20180410_091924.jpg

time stamp 1526546180 (Thursday May 17, 2018 10:36:20 (am)). We also assume
that this time stamp could be the end of the raid. We reason that, the very last thing
the criminals would do is attempt to hide evidence of their involvement by setting
the lab on fire.

3.5 iSmart Alarm

Home alarm system that can be setup by anyone. It secures any doors or windows
with the contact sensors. It uses a centralized brain/control system which is called
Cube One. The Cube One has built-in Siren to ward of intruders. The system has

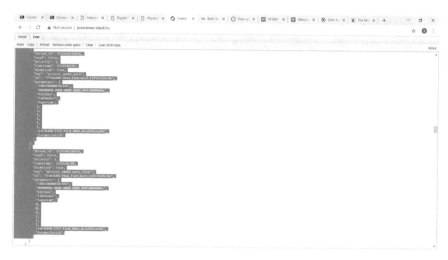

Fig. 4 Nest protect alert /img_blk0_sda.bin/vol_vol21/app/com.nest.android/cache/cache

Motion sensors that will trigger alerts to your phone during the day. The system has a variety of cameras. The cameras can be controlled remotely by your phone and/or remote controllers. One camera offers a pan and tilt feature. This system was used to alert the drug runners of any intruders or police.

3.5.1 Device Features

- Most iSmart alarms include: Includes 1 CubeOne™ (with built-in 110 dB siren), 4 Contact Sensors, 3 Motion Sensor, and 3 Remote Control Tags
- Emergency alerts to include phone call, text message, app push notification or email.
- IFTTT compatible.

3.5.2 Forensic Analysis

We believe the iSmartAlarm has 3 users: Jesse Pinkman, D. Pandana, and S. Varga. We assume that operator pandadodu is D. Pandana and TheBoss may be S. Varga. If we adjust the times for UTC +2, then we could see that Jesse armed his device at 10:22:22 and 8 seconds later TheBoss disarms the security system. Our best assumption for this, is that Jesse armed the system via his phone because throughout the day he never set the device to Home, which would indicate him being in the lab. TheBoss set the device to Home indicating they were at the lab at 10:34:17 and Pandadodu disarmed the device at 10:34:31. These events happened minutes before the police were alerted, leaving us to believe that S. Varga and Pandadodu may both have been involved in the raid (Figs. 5 and 6).

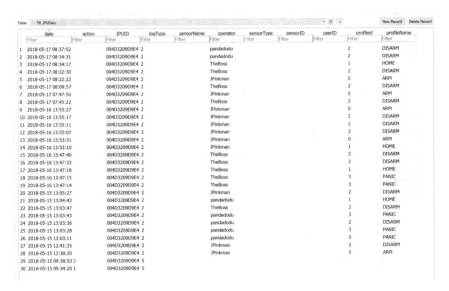

Fig. 5 (iSmartAlarm.db) File /img_blk0_sda.bin/vol_vol21/data/iSA.common/databases/iSmart-Alarm.DB

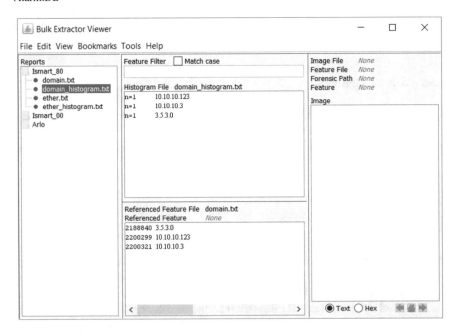

Fig. 6 iSmartAlarm and other associated IPs bulk extractor view of ISmart_80.img

We've run the iSmartAlarm files through Bulk Extractor and have found what we believe to be the IP addresses of the 3 devices that consist of the iSmartAlarm system. We also found the IPs of devices the iSmartAlarm may have been previously connected to. The ISmart-00 file had the exact same IP addresses on it

3.6 Wink Hub

The Wink Hub is a device that controls and manages everything in your home. The Wink Hub can turn lights on and off inside or outside your home. It can remotely control your home's thermostat, garage door, ceiling fans, outlets, locking and unlocking of your front door. The Winhub can control your home appliances, and ZIGBEE. The Wink Hub can be accessed and controlled from your personal Apple or Android phone.

3.7 Qbee Camera

This camera too has a pan and tilt feature that can be remotely controlled by a smartphone. It could be used to monitor your pet's activities during the day, even letting you pan the room to see what your pet is up to while you are at work. This camera could have been used to remotely monitor activity in the room. We wanted to use the Temperature and Humidity reading to prove the time of the fire attack on this house.

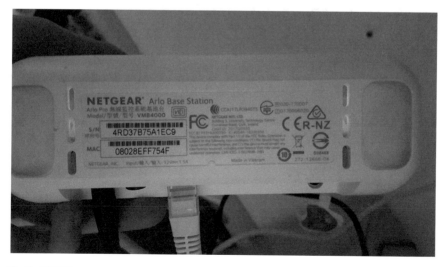

Fig. 7 (NETGEAR) Arlo base station MAC address /img_blk0_sda.bin/vol_vol21/media/0/DCIM/ Camera 20180410091838.jpg

Fig. 8 (SamsungE_25) Wink hub MAC address /img_blk0_sda.bin/vol_vol21/media/0/DCIM/ Camera 20180410_091838.jpg

Fig. 9 (NestLabs) Nest camera MAC address /img_blk0_sda.bin/vol_vol21/media/0/DCIM/ Camera 201800410_091920.jpg

3.8 Samsung S6 Edge

We ran the Samsung's files through Autospy and found data pertinent. The data found here is very important to answering the Attorney General's questions.

Fig. 10 (AskeyCom) QBee
camera MAC address
/img_blk0_sda.bin/vol_vol21/
media/0/DCIM/Screenshots
Screenshot_20180502-
132904.png

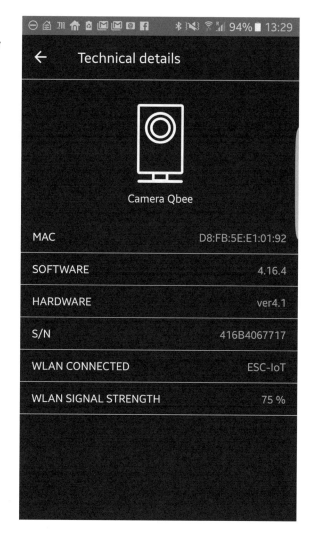

3.8.1 Camera Evidence

The images found on the phone's camera that give us the MAC addresses. We were later able to link these to IP Addresses found through the PCAP file (Figs. 7, 8, 9, 10 and 11).

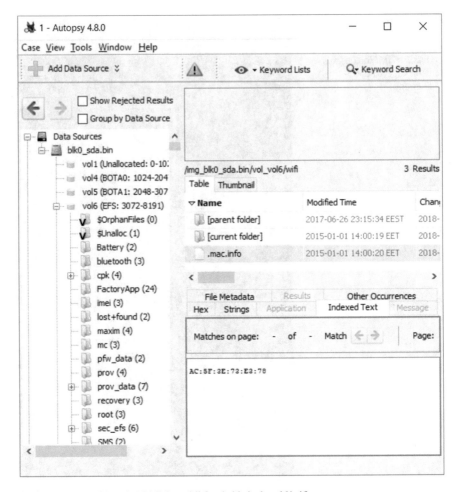

Fig. 11 Samsung Phone's MAC /img_blk0_sda.bin/vol_vol6/wifi

4 Additional Findings and Assumptions

4.1 Possible Qbee Camera Disable Times

We searched for keyword "qbee" in Autopsy and found some .xml files that contain information about the Qbee Camera. We extracted the files and put them into the XML viewer and managed to find information like the last active time and the beginning idle time of the QBee camera. However, we couldn't find the right time converter to convert the time into a readable format. We also found another file called "appops.xml", extracted and put it into the XML viewer, this time we are able to convert the timestamp (Fig. 12).

Fig. 12   Qbee Camera last active times /img_blk0_sda.bin/vol_vol2//$CarvedFiles/f0320104.xml
/img_blk0_sda.bin/vol_vol21/system/appops.xml

First timestamp: 1526384194498 converts to Tuesday May 15, 2018 13:36:34
Second timestamp: 1526564163886 converts to Thursday May 17, 2018 15:36:04
Third timestamp: 1526564163886 converts to Thursday May 17, 2018 15:36:04

4.2   Nest Camera Deleted Footage

According to one of Jesse's emails, his Nest Camera was removed from his account, which would end up deleting all footage and forgetting familiar faces. "Heads-Up. Office (LabCam) camera has been removed from your account. Its serial number is 18b43061c9ef. Its video history has been deleted and familiar faces will remain unless this is your last subscribed camera." This means that if Pandana or Varga went into the lab they would be strangers to the camera and Jesse would be alerted. However, this event occurred on 5/15; 2 days is sufficient time to re-add them as familiar faces. We do believe though, that this could have been the early stages of the raid. Perhaps someone was trying to disable all of the cameras (Fig. 13).

Fig. 13 Nest camera's history deleted

4.2.1   Nest Camera Disabled

"last_disconnect_time": 1526551423000,
"cvr_enrolled": "5-days",
"last_connect_time": 1526552218000,
"activation_time": 1526289799580,
"websocket_nexustalk_host":   "oculus294-eu1.dropcam.com:80",   "model":
"Nest Cam",
"description": "",
"direct_nexustalk_host": "oculus294-eu1.dropcam.com",
"audio_input_enabled": true,
"download_host": "oculus294-eu1.dropcam.com:80",
"recorded_stream_host": "oculus294-eu1.dropcam.com:1935",
"mac_address": "18b43061c9ef",
"structure_id": "a3e36480-5757-11e8-80bb-0e2d565eed46",
"last_disconnect_reason": "no_reason",
"software_version": "217-610040",
"wwn_stream_host": "stream-eu1-delta.dropcam.com",
"live_stream_host": "oculus294-eu1.dropcam.com:1935",

According to this image, the Nest Camera was disabled from 10:03:43 AM to 10:16:58 AM. This could show that someone was trying to disable all the cameras before the raid happened (Fig. 14).

Last disconnect time: GMT: Thursday, May 17, 2018 10:03:43 AM

Last connect time: GMT: Thursday, May 17, 2018 10:16:58 AM

Fig. 14 Nest camera disabled

4.3   Assumptions Based off Amazon Echo Data

Based off the above evidence, we conclude that Jesse Pinkman was not at the lab during the raid, rather, he left approximately 10 minutes before (Fig. 15).

Fig. 15   Jesse Pinkman arming iSmartAlarm

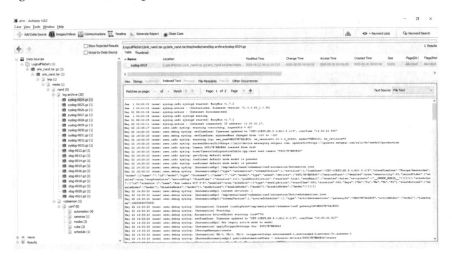

Fig. 16   Arlo camera tar.gz file

4.4 Arlo Dead End

We found nothing special from this file, just some syslog files that gave us information about the Arlo's IP address. And the earliest syslog file that can be found here is on May 22, which is after the raid (Fig. 16).

Fig. 17 Wink hub tar.gz files

4.5 Wink Hub Dead End

Again, we found nothing pertaining to the raid. Most of the files found here were empty. 3 log files appeared to have occurred on May 17, the day of the raid, but we didn't find anything relevant (Fig. 17).

Part II
Technical Case Studies

Alexa/Google Home Forensics

Sean Tristan, Shally Sharma and Robert Gonzalez

Abstract Amazon Alexa and Google Home are a potential source of digital evidence, partly because of their popularity. In additional, these devices are also always-on and listening to our day-to day conversations. Therefore, in this chapter we study Amazon Alexa and Google Home, the two most popular speakers. We will discuss how the voice service of these two devices work, what all information these devices potentially record, and where this information stored. Then, we will compare the accessibility of data stored locally on each device as well as data stored remotely associated with the specific device. The goal of the research is to see if the information collected by these smart devices could potentially have substantial value in a forensics investigation.

1 Introduction

Internet of Things (IoT) devices are becoming more popular day by day with smart homes becoming commonplace in the foreseeable future. With this trend in mind, companies like Google and Amazon are capitalizing on their previously established market base, enhancing the user experience with their intelligent voice-activated cloud-based services, Google Home and Amazon Echo respectively. These devices can fulfil a variety of services such as playing music, providing news briefs, ordering food, and turning on lights, just by simple voice commands from the user. These smart speakers connect to the Wi-Fi network and can also be controlled via an application on the smartphone [12].

S. Tristan · S. Sharma · R. Gonzalez (✉)
Department of Information System and Cyber Security,
University of Texas at San Antonio, San Antonio, TX, USA
e-mail: robert.gonzalez@my.utsa.edu

S. Tristan
e-mail: sean.tristan@my.utsa.edu

S. Sharma
e-mail: sharmashally030@gmail.com

© Springer Nature Switzerland AG 2020
X. Zhang and K.-K. R. Choo (eds.), *Digital Forensic Education*,
Studies in Big Data 61, https://doi.org/10.1007/978-3-030-23547-5_7

These smart devices were created to seamlessly enhance our lives and automate mundane tasks we have to endure every day. These devices over the years have made their way into millions of homes across the globe. Approximately 8.2 million Echo devices are owned in the United States alone which is a 60% increase from 2016. As so many smart homes our now being installed in people homes and apartments the questions regarding data privacy become more essential than ever. Are these smart devices always listening? If so, how much of our daily conversations are being recorded? If they are being recorded how much of this information is being stored in a backend database somewhere? These types of questions are what consumers are worried about and some of these FAQ's are the reason why some people have declined purchasing a smart device for the fear of even more invasion into their personal lives [12].

2 Background

When Amazon first rolled out the Echo back in 2014 it was first introduced as a smart speaker which would allow people to control their music with only their voice. Soon after the initial release Amazon went to work on improving the virtual agent that powers the Echo and its steadily got more and more capable of additional features. Amazon Echo evolved from simply a smart speaker to the standard for home automation through voice control. Now there are more than 12,000 products that have Alexa skills integrated into their platforms [15].

2.1 1966–1967: Echo IV/Kitchen Computer

The Echo IV was the first "smart device", although this device was never commercially sold. The Echo IV had similar functionality as some of the smart homes today however, it wasn't nearly as sophisticated. The device could create shopping lists, control the temperature, and turn appliances on and off. The Echo IV was comprised of four large (6' × 2' × 6') cabinets weighing almost 800 lbs [13]. This device included a central processing unit that was made up of surplus circuit modules from a Westinghouse Prodac-IV. Jim Sutherland requested permission from his employer at the time, Westinghouse, and took these modules home and was able to design the Echo IV in less than a year (Fig. 1).

2.2 1991: Gerontechnology

Gerontechnology combined with technology intends to improve the lives of senior citizens. In the early 1990s there was a lot of new research and technology in this

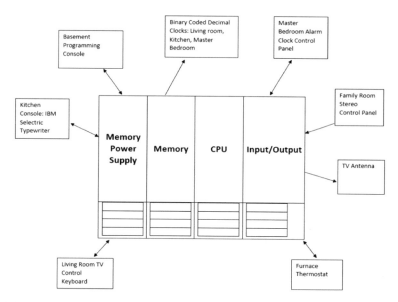

Fig. 1 ECHO IV system diagram

division. An example of such technology is the commercial "I've fallen and can't get up" [5].

2.3   1998—Early 2000s: Smart Homes

In the early 2000s is when home automation really began to increase in popularity. During this time many different technologies and patents began to emerge and functionality also took a leap forward. Smart homes became more affordable and with more and more homes becoming smart homes a viable market began to emerge.

One of the first problems with Alexa was how to overcome the latency issue or how long it would take Alexa to respond to a question. According to Business Insider the average latency of existing smart speakers was around 2.5–3 s so the goal during implementation was to get Alexa to respond in under 2 s. However, after proposing the initial target to Jeff Bezos (CEO of Amazon) he countered with the target latency of 1 s. The key to reaching such a low latency interval was to collect as much data as possible in order to automate some of the more common/frequent tasks and apply them to improve the product. After thousands of trial tests and data analysis the team at Amazon was able to work with speech scientist and eventually achieve the goal of below 1.5 s latency which was far exceeding any of the competitors in the market at the time [8].

3   Smart Speaker Market

In 2014, Amazon released its Echo—a cylinder-shaped device, only available to its Amazon Prime members. This device was made available to the general public in 2015. After its success, a smaller version called Echo Dot was released in 2016. Both Echo and Echo Dot are powered by a cloud-based voice service, Alexa. In May 2016, Google announced its own device, Google Home, which was released in November, 2016. The following year, Google released its less expensive variant—the Google Home Mini [9].

Applications that connect smart speakers to a variety of platforms and devices are used to initiate more complex tasks such as adding items to to-do lists or shopping carts. These apps can also be used to control music and other smart home devices. They can also set timers/alarms and, when away from home, allow one to control lights and doors from a remote location. All commands and responses are synced with an app, regardless of the path the task was initiated with. For example, if the Echo and Echo Dot are connected with the same account/app, they will remain in sync [12].

4   How Amazon Alexa Works?

So how do these smart speakers work? All the "magic" happens in their voice recognition software. Once a smart speaker is ready for use, it listens to all the sounds and waits for its respective wake command—'Hey Google' for Google home and 'Alexa' for Amazon Echo, before it gets into action mode. Once the device detects its wake command, it starts recording speech and this speech file is sent to the respective voice recognition service in the cloud. The voice recognition service interprets the speech and sends back the response to the smart speaker. It uses algorithms to become more familiar with one's use of words and individual speech patterns [14].

Alexa has certain skill set that gives it the ability to respond and initiate tasks for its users. These skills are formulated by developers and are enabled by the user to bring it in action. Therefore, it can only perform functions that are within the range of these skills that have been added by its developers. If the user requests for something that is beyond this range, Alexa is unable to process it. Amazon Alexa voice service, a force behind Echo and Echo Dot, has more than 10,000 skills available to create a more personalized experience for users as per their interests and needs [12].

These skills are divided into three major categories and they are custom skills, smart home skills, and flash briefing skills. Just like each category requires different level of work from its developer, similarly each type requires a different level of effort or command path from the user when invoking these skills [12]. Let's look at these skills individually:

1. **Custom skills**—These are the skills that are developed from scratch and requires to define some important components such as intent (skill's functionality), Utterances (spoken words by the user for a particular action), Invocation name (desired name for the skill), and cloud-based service (Internet accessible endpoint for a skill). Even though it is a difficult and complicated skill to create, it is the most flexible that can deal with variety of requests [12].

2. **Smart home skills**—These skills are the ones that gives their users the ability to control their internet of things (IoT). It very simple to use these skills as long as they are enabled, and the devices are connected and synced. Once it is done, user simply needs to speak to its smart speaker to initiate the task. It is easy to build such skills for the developers as they do not require to build everything from scratch. These skills are built by using Amazon's Smart Home Skills Application Program Interface (API) where interaction models are already available for use [12].

3. **Flash briefing skill**—This is the easiest skill to create for developers. These skills help provide contents such as news in summary to its users. This is called flash briefing and users need to enable these content skills to use this feature [12].

As the popularity of these smart speakers is growing day-by-day, many companies are developing skills and devices that are compatible with Alexa and other such speakers to take advantage of their presence. Companies such as Domino's Pizza, and Capital One are one of those companies which have invested their resources in creating custom skills to advance their reach as a business [12].

5 Market Share of Echo and Home

Studies show that, even though Google entered the market almost two years later, it is catching up to Amazon. As per a research firm Canalys in May 2018, Google's smart speaker sales has skyrocketed to 483% and has outsell Amazon. According to eMarketer, the market for smart speaker as a whole is also growing dramatically by 48% annually [9].

According to a report published in August 2018 by a research group called CIRP (Consumer Intelligence Research Partners), smart speakers have reached a count of 50 million in the U.S. market. This market was invented by Amazon in the year 2015 and it is still a clear leader with a market share of 70%, whereas Google has a 24% market share. The remaining 6% belongs to the Apple HomePod, a new smart speaker launched by Apple. The graph below depicts the market share of Amazon, Google and Apple from mid-2017 to mid-2018 [9].

According to CIRP, Apple's HomePod will have a hard time competing with Amazon and Google. Even though it does have a small share of the market but still both Amazon and Google have already created a niche and are far less expensive when compared to HomePod. Apple is selling it at approximately $350 whereas

Home and Alexa's mini version costs less than $50 and is even cheaper if on sale. The good news is that the market is growing rapidly and as per researches, the smart speaker owners are loyal to the company they buy from because they buy by voice [9].

6 Why Are They Listening?

Amazon Echo collects variety of data to give personalized user experience to its users which is saved on the amazon server and they are as follows [12]:

- Skills that user enables
- All Questions and requests made by the user
- All changes made by the user
- All responses that Echo gives to the user
- Subscriber information that includes shipping/billing address and payment method
- Information provided by linked accounts/third party services.

As per Amazon, Google, and Apple the data that is collected and stored on their servers (which could be located anywhere in the world irrespective of the country the user is in) is used to analyze user trends and helps these companies to provide more customized services and features to its users. Another reason that the data collection is significant is because it is required for certain skills to work the device must have access to that information. For example, when a user requests something that is linked to third party service such as Spotify, the smart speaker must have access to user's Spotify account to obtain to fulfil user's request [12].

Smart speaker's voice recognition capability is built with artificial intelligences which is capable of natural learning. This capability is constantly trying to improve itself by using the data that it is storing and is learning from it as the more the speech recognition engine takes in, the better it gets in recognizing and understanding different voices, accent, and dialects. This improves the user experience with the device and makes the response more effective and efficient [12].

7 Are They Always Listening?

As the popularity of these smart speakers continues to grow, one wonders how safe these devices are when it comes to user privacy. There have been many concerns raised since the advent of these devices. Are these products always on and actively listening to the conversations taking place in households? Is the information is being recorded? If yes, then what is this information being used for?

Amazon has addressed these concerns raised towards its smart speaker Amazon Echo and has provided details on how their device works. As per Amazon, Echo is always active and listening for its wake work. It does not pay attention to the words spoken around it unless it hears the wake word. That is when it becomes attentive and starts recording. This recording is processed in the cloud and generates a response for the user. According to Amazon, the recorded speech is saved on Amazon serve to provide personalized experience to the customer [12].

Another significant concern that has been raised is whether the collected information can be used by law enforcement. If so, does it possess any evidentiary value? The study conducted by Douglas A. Orr and Laura Sanchez studied to determine whether the data generated by Amazon Echo and Alexa Voice service app could be of any evidentiary value. They studied and analyzed the data located online at alexa.amazon.com, the folder structure of Alexa app, and in mobile device's cache. They used two methodology to acquire and analyze the data: manual extraction and logical backup extraction [12].

After the extraction part, they analysis the collected data in two ways. First they analyzed the data manually and later used forensics tool Blacklight, version 2016.3.1. They were able to retrieve and analyze a lot of information from History, Settings, clips of recorded speech, and Home cards. The results of their analysis suggests that Amazon Echo does possess data which has evidentiary value. Data such as timestamps and the content of recorded data can be used as potential evidence that can help the law enforcement in their investigation process. The data collected by these devices can be used along with other source of evidence to prove or refute a crime [12].

8 Alexa Forensics—Data Access and Retrieval

Amazon's Alexa voice-assistant system, with its variants *Echo* and *Echo Dot*, have been referred to as "a human-life's black box" [4]. The amount of highly-personal data which can be collected and processed through these devices is virtually unlimited, given that they interface with Amazon cloud-based storage system. The challenge for a forensic investigator lies in the ability to access to and retrieve such data. A very limited amount of user data is stored on the device itself in its flash memory hardware. The remainder of must be accessed through cloud storage [3]. These devices can also be networked with cell phones, tablets, laptops, and various other IoT smart devices and applications but must be routed through the user's account within the Amazon's cloud system.

The staff at iFixit physically disassembled Amazon's *Echo* unit and found that the device houses a small motherboard which includes the following key components:

- Texas Instruments DM3725CUS100 Digital Media Processor
- Samsung K4X2G323PD-8GD8 256 MB LPDDR1 RAM
- SanDisk SDIN7DP2-4G 4 GB iNAND Ultra Flash Memory
- Qualcomm Atheros QCA6234X-AM2D Wi-Fi and Bluetooth Module
- Texas Instruments TPS65910A1 Integrated Power Management IC [6].

The iFixit staff also found that reassembly was very difficult and it must be noted that this action is beyond the scope of our project. Awareness of the internal components, however, does offer insight into avenues of approach for data access and retrieval. Three such avenues are offered by Clinton, Cook, and Banik, "These are, in order of perceived difficulty (easiest first): the SD card pinout, an eMMC style root, and finally JTAG" [3]. Barring the use of a paired device such as a cell phone with access to a user's Alexa cloud account, our group would most likely follow the first approach.

Amazon uses a low-latency cloud-based BI platform for automating the retrieval of data through queries. Similar to other cloud services, Alexa uses predefined API's to transceive data however, these API's are not publicly available. Many forensic researchers have been performing client-based forensics but suggest examining the cloud native module for additional information since a majority of data is stored in temporary caches [4]. However, there are some limitations involving cloud analysis. To access the cloud, you must first have an account with Amazon and it is also difficult to retrieve deleted data that is stored in the cloud. According to the article, after the research was conducted a proof of concept was created for cloud-based IoT environments "CIFT: Cloud-based IoT Forensics Toolkit" which can acquire cloud native artifacts from Alexa using API's that allow client-based artifacts to be compared to the web-based application [4]. As seen below the CIFT design has four major components which are the UIM (User Interface Module), CNM (Cloud Native Module), CCM (Companion Client Module), and DPM (Data Processing Module).

Android and IOS devices can access Alexa using the official mobile application developed by Amazon. With the Android system, it uses two SQLite files: map_data_storage.db and DataStore.db, the first contains token information about the current user logged in [4]. Deleted records can also be found in unused areas of SQLite database and its journal file. These files contain information such as to-do lists and shopping lists. In the iOS version of the application SQLite has a file titled LocalData.sqlite which is similar to the file found for android and also contains list information. The article states that little information is actually stored on the companion devices however, XML or PLIST files also exist for storing preferences they were not used in the study [4].

9   Google Home Forensics—Data Access and Retrieval

Google Home's voice-assistant, with its variants Google Home Mini and Google Home Max, were originally marketed simply as "intelligent speakers" but the addition of numerous apps and features associated with these devices have since made it clear that Google's intention is to have its devices compete directly with Amazon Alexa system. Similar to Alexa, Google Home is networked through Google's cloud infrastructure and can be paired with cell phones, tablets, laptops, and other smart devices.

Phill Moore, a digital forensics analyst from Australia, in his research used two approaches for data acquisition to examine the data on cloud for Google Home. One was a manual approach, the other involved the use of tools such as Magnet Forensics and Axiom. In the manual cloud acquisition approach, he gained access to "My Activity" and was able to obtain information on spoken words, responses, audio, and location information [10].

He followed a four-step process in his research in order to gain access to the data. The first step was to get physical acquisition of the Android phone connected to the Google home device. The second step involved identifying the Google Account and Cloud ID from Home Graph. The third, necessitated that one query a home device to obtain the cloud ID and Bluetooth data. The last and final step involved actually downloading data from the cloud [10].

Moore found that he was able to retrieve a significant amount of information about the owner of a device. He was able to find the name of the person in relation to the room where the device was based. He could figure out the person's activities with the help of calendar entries. He could also determine the previous location as it was set as his home location, the potential last date/time based on a Bluetooth connection, and a person's interest in places based on queries regarding weather in certain locations [10].

Unfortunately, not as much information is extent regarding digital forensic studies on Google Home devices when compare to those on Amazon Alexa devices. Moore, however, also revealed at a 2018 SANS Digital Forensics conference that he had created a python script named "Homespeak" which he had developed in order to retrieve information from Google Home devices connected to their cloud accounts [11].

The staff at iFixit physically disassembled a Google Home unit and found that the device houses a small motherboard which includes the following key components:

- Marvell 88DE3006 Armada 1500 Mini Plus dual-core ARM Cortex-A7 media processor
- Toshiba TC58NVG1S3HBA16 256 MB NAND flash
- Marvell Avastar 88W8887 WLAN/BT/NFC SoC
- Texas Instruments TAS5720 audio amplifier
- Samsung K4B4G16 512 MB B-Die DDR3 SDRAM [7].

10 Hardware Differences—Implications for Data Access and Retrieval

A comparison of hardware from both systems shows some notable differences. The Echo unit includes onboard flash memory of 4 GB versus 256 MB for the Google Home unit. However, the Echo device featured onboard RAM of 256 MB versus 512 MB for the Google Home device. The greater flash memory capacity for the Echo unit would mean that digital forensic investigators should expect to access more local data from these units. The Google Home device with its higher RAM capacity is also a target for investigation but data access and retrieval from RAM is significantly more challenging as the unit must be running and care must be taken to prevent or at least minimize the introduction of data artifacts during the forensic analysis procedure.

11   Data Retrieval Methodology—A Seven-Phase Framework for Smart Speakers

Our team sought to observed a seven-phase framework which has been proposed as a Home Automation Systems (HAS) forensics framework and covers the main steps required for gathering digital evidence [2]. The creators of this framework state that all seven steps are not required for every investigation and that they should be used as ay the investigating team's discretion. This team found this to be the case, especially since physical access to internal local data storage was determined to be beyond the scope of our project for the larger, more expensive devices. In general, these phases can serve as a great reference where the smart speakers are involved in the investigation process.

The following is a summary of each phase and how they relate to the forensic investigation of smart speakers:

- *Phase 1*:
  Preparation off-site—This requires having a forensic investigator with appropriate expertise who is well versed with processes and procedures relating to the investigation. This phase also includes the availability of required tools for data identification and acquisition. This phase a required phase when we are dealing with smart speaker's forensics.

- *Phase 2*:
  Search for a home automation system on-site—Here the investigation team needs to locate the physical devices that can indicate the presence of Home Automation System. A physical search of the scene can help locate the smart speakers. This phase can be a helpful tool if such devices are not located in plain sight.

- *Phase 3*:
  Preserve the HAS 'as is', wherever possible—This is the phase where the forensics team is required to find and contact the service provider involved, if any. This is also relevant for smart speakers as most of the data is stored on the cloud. Therefore, the digital forensics team is required to contact the provider by following proper channels to retrieve and preserve the data to ensure evidence admissibility.

- *Phase 4*:
  Understanding the specific home automation system—This is where a global picture of the system is created such as network topology and a detailed overview is prepared which includes the technology, Manufacturer, Configurations, Connectivity, and command and control information.

- *Phase 5*:
  Check Security level—This is important for the investigation process to know the security measures in place on the system in question. This includes if the system has enabled access controls, user lists and their rights, and other security measures.

- *Phase 6*:
  Locate and acquire evidential data—This phase is an important phase once the potential source of evidence if identified. This is where the evidential data is acquired with the help of tools and techniques relevant for the particular device. Smart speakers have most of its data stored on the cloud servers and very little is present in the device itself. Therefore, the investigative team is the required to use the tools and process for data acquisition.

- *Phase 7*:
  Process/analyze seized data—In this phase the seized data is analyzed and information is recovered from it that can be useful in the investigation process. It is a very significant phase used in smart speaker data analysis.

12 Investigation/Findings

Amazon Echo (1st Generation) Device

The Amazon Echo (1st Generation) device does not have accessible external physical ports. The only connection is a jack for the power cable. Barring prying open the device case which is not designed to be disassembled and reassembled, the only option for physically non-destructive data retrieval is through a connection to the device through Amazon's cloud data storage. Amazon's "Alexa app" is designed to enable this access. An initial attempt to retrieve data via a Blackberry Classic cell phone failed though since a third-party application failed to install. Amazon does not offer a version of its Alexa application that is compatible with the Blackberry OS. Available versions are only compatible with iOS or Android phones (Fig. 2).

A second attempt was made with a ZTE cell phone running Android 6.0.1. The Alexa application was installed on this phone. The application featured a multi-step process which enabled the Echo device to connect to the internet. The first step involved logging into the Amazon server . This required knowledge of the Amazon

Fig. 2 Blackberry classic (Blackberry OS) cell phone and Amazon Echo (1st generation) device

account holder's username and password. The Echo device was plugged in and a greeting announcement was made through its speakers as it began to scan for a network connection. The device failed to connect to its previous network and an announcement was made about it not being able to connect. Using the application, the phone began to broadcast a signal to the Echo device. A button on top of the device was pressed to put it into listen mode. The device was able to detect the phone and the application displayed a message prompting one to search the phone's Wi-Fi networks for a temporary broadcast from the device. The phone was connected to this temporary network and the application allowed one to enter the network password information to allow the Echo device to establish a direct internet connection.

Once this connection was established a serious of voice commands were issued to test the device. Next, the Alexa phone application was used to access details about the Echo device including software version (613505920), serial number (90F00718626208C0), and MAC address (50:F5:DA:24:51:46). Next an attempt was made to retrieve usage history from the device but the results only displayed the voice commands issued to test the device connection. No prior data was available. Either the previous user deleted the command history from their account or a security measure prevented access to this history (Figs. 3 and 4).

Fig. 3 ZTE (Android 6.0.1 OS) cell phone and Amazon Echo (1st generation) device

Fig. 4 ZTE phone and Amazon Echo information and Amazon Echo Dot (2nd generation) device

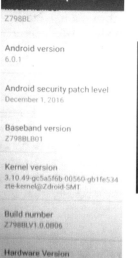

Amazon Echo Dot (2nd Generation) Device

The Amazon Echo Dot (2nd Generation) device is a personal assistant that gives your existing speakers Alexa's eyes and ears so that you can speak to play music or check the weather. The one question we are interested is what Alexa is made of and what makes it operate. What can we expect to see inside the Echo Dot? Alexa can speak and listen so there is most likely going to be a speaker and microphone. The Echo does have a USB connection but I was unable to pull any information from it by plugging it into my computer. The other port on the back of the Alexa allows you to hook up an additional speaker via auxiliary cord. I asked Alexa what she is made of. Alexa's response was "I am made from various plastics and metals" so I decided to find out for myself.

Phase 1:

At first, we noticed there were no screws anywhere around the Dot so we took a razor to the back of the Echo Dot and started peeling back the adhesive. Once we were able to peel the adhesive off of the non-slip pad we had to remove the plastic covering the screws.

Phase 2:

Under the plastic was (4) T8 Torx screws that were threaded all the way through the Echo Dot. Once we removed all the screws the body of the Dot slid right off. When we lifted off the base we noticed a large piece of plastic attached to the base which housed the speaker for the Dot. The plastic is formed around the speaker to amplify the sound of the device.

Phase 3:

Besides the plastic piece the rest of the device was held together by a ribbon cable that was connecting the motherboard to the bottom printed circuit board with the LEDs and microphones on it. We had to disconnect the ribbon cable to get the other pieces to separate as seen below (Fig. 5).

Phase 4:

Alexa was right essentially she is made up of layers of plastic and metal and two Printed Circuit Boards (PCB). The first board that sits right under the buttons of the Dot, it contains seven microphones along the outside and one in the center. The 4 chips in the middle are analog to digital converters (ADCs) which are translators between our speech and a digital voice that Alexa can understand. The Digital to Analog Converter (DAC) on the motherboard then takes that digital voice and makes it analog again and is what we hear coming from the Alexa Dot (Fig. 6).

Phase 5:

On the other side of the PCB there is 4 tactile dome switches the switch that belongs to the action button has a light sensor on it that adjusts the brightness of the LEDs according to the brightness of the room (Fig. 7).

Fig. 5 Ribbon cable

Fig. 6 Four analog to digital converter and microphones

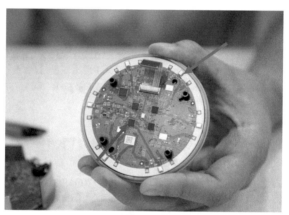

Fig. 7 Four Tactile Dome switches

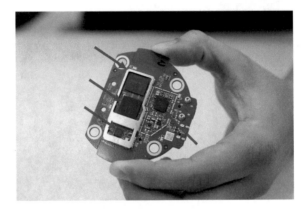

Fig. 8 - 64-bit quad-core CPU and mediatek chips

Phase 6:

On the motherboard there is a Mediatek power management IC which converts voltage coming from a USB. We also found a system mod chip wireless module that houses all the Wi-Fi, Bluetooth, and GPS capabilities. In the center there is a 64-bit quad-core arm cortex a53 CPU which is a Mediatek chip, the last chip is a combination of both flash and RAM for processing [7] (Figs. 8, 9).

13   Wireshark Capture (Echo Plus)

Since the Echo Dot has been setup to run on my home network. I installed Wireshark to see what kind of traffic is generated from the device? We filtered the traffic based on the MAC address of the Echo that was found in the GUI for the device. From the packet capture we can see that a majority of the traffic to and from the Echo Dot uses TLSv1.2 encryption with certificate validation. Since it is CA validated it prevented us from attempting a MITM (Man-in-the-middle) attack on the device. I decided to plug in the Echo Plus and start it in the setup mode to get it connected to my phone and my current network (PingSlayer). The Echo Plus created a Wi-Fi access point called "Amazon-13v" which I had to connect to in order to get it linked to my Amazon account and home network. While Wireshark was running I was consistently asking Alexa questions to see what kind of traffic would be generated. I noticed a lot of DNS requests to domains called pindorama.amazon.com and pin.amazon.com. These domains I believe to be the Echo's web-based control sites (Figs. 10 and 11).

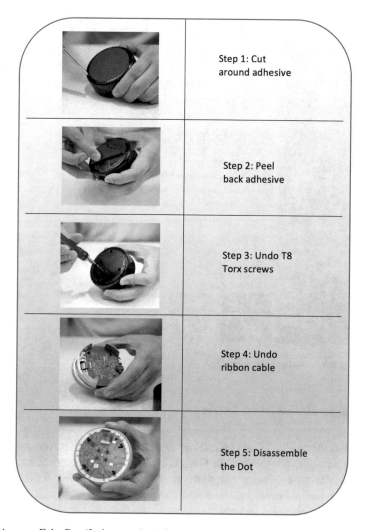

Fig. 9 Amazon Echo Dot (2nd generation) disassembly

No.	Time	Source	Destination	Protocol	Length	Info
4617	281.772486	2600:1700:1400:ba40:2c75:321a:7eb1:80d0	googleapis.l.google.com	GQUIC	98	Payload (Encrypted), PKN: 7, CID: 15135563920717940337
4618	281.777893	2600:1700:1400:ba40:2c75:321a:7eb1:80d0	67.1.168.192.in-addr.arpa	DNS	95	Standard query 0x7096 A pins.amazon.com
4619	281.777940	2600:1700:1400:ba40:2c75:321a:7eb1:80d0	67.1.168.192.in-addr.arpa	DNS	95	Standard query 0x87d2 AAAA pins.amazon.com
4620	281.793423	67.1.168.192.in-addr.arpa	2600:1700:1400:ba40:2c75:321a:7eb1:80d0	DNS	111	Standard query response 0x7096 A pins.amazon.com A 23.23.78.1
4621	281.798234	2600:1700:1400:ba40:2c75:321a:7eb1:80d0	googleapis.l.google.com	GQUIC	98	Payload (Encrypted), PKN: 8, CID: 15135563920717940337
4622	281.826301	67.1.168.192.in-addr.arpa	2600:1700:1400:ba40:2c75:321a:7eb1:80d0	DNS	143	Standard query response 0x87d2 AAAA pins.amazon.com SOA ns-924.amazon.com

Fig. 10 Wireshark capture

Fig. 11 Amazon Alexa application display

14 Google Home Mini Device

Unlike the Amazon Echo (1st Generation) device which only has a connection jack for simply powering the device, the Google Home Mini features a female micro-B USB port to charge the device which may potentially be an avenue for data retrieval though such a method is not has not been supported or published by Google. Similar to the Amazon Echo (1st Generation) device, the Google Home Mini is also not designed to be disassembled and reassembled so the only option for physically non-destructive data retrieval is through a connection to the device through Google's data cloud storage. The Google Home application, available for iOS and Android mobile device, is designed to enable this access (Fig. 12).

The initial attempt to connect to the device was made at the Main Campus of the University of Texas at San Antonio (UTSA). The Google Home application was successfully installed on the same ZTE mobile phone used to connect with the Amazon Echo (1st Generation) device. Unfortunately, this attempt failed as the wireless network connection would be lost during the setup process. A representative from Google's online support was contacted and she explained via chat the problem was due to the fact that the Google Home application did not support the use of public network connections which require the use of a captive portal. This commonly involves a login page displayed through a browser where one is required to enter their

Fig. 12 Equipment pair: ZTE (Android 6.0.1 OS) cell phone and Google home mini device

credentials such as a username and password. The captive portal acts as a gateway before one can access the internet on a network [1]. UTSA's Rowdy Student Guest network is a public network which uses a captive portal where one must enter a student ID and password.

A second connection attempt was made at a team member's network at home. This attempt was successful. During the setup process, network information including the password for login authentication was ported over by the Google Home application to the Google Home Mini device.

The Google Home application provided the following basic device information:

System firmware version: 139856
Cast firmware version: 1.36.139856
Language: English (United States)
Country code: US
MAC address: 20:DF:B9:0D:FF:8F
IP Address: 192.168.1.81.

Most relevant to our data retrieval effort:

Without access to their username and password, the Google Home application also revealed information on the account of the person who previously used the device as part of its "Linked account(s)" data. This account information included a thumbnail image and name both of which may be used to identify the individual who owned this device. Note that the image resolution on the right is much better on the phone and has deliberately been blurred here. This combined information can serve as a basis for further investigation on the likely previous owner of the device (Fig. 13).

Fig. 13 Google home mini device and previous user account (blurred)

15 Conclusion

Smart devices are now another means of digital data collection and can be increasingly relevant in a digital forensics investigation as they become more widespread. Their stored data can come in many different formats and can contain pertinent information coming from human and or system interactions. Home Automation Systems (HAS) can be any device with access to the internet that allows homeowners to easily control the device via smart phone or voice activation. Data that is stored on these devices such as audio, images, or video could be collected and be used against a suspect in a crime investigation to reconstruct a timeline of events. The more familiar forensic investigators become with these devices the easier it will be to understand how they could be involved in a criminal investigation.

There are a number of challenges, however, when it comes to forensics and smart homes. For instance, the need for standardization is a big challenge. Each smart home system on the market currently has its own unique set of subsystems with a variety of different applications and features. This, as well as differences in native hardware storage architecture, makes it difficult for software engineers to create forensic tool kits that can be used to access and retrieve data from these smart devices. Presently, many more resources exist for accessing and retrieving data from Amazon Alexa than from Google Home devices. Hopefully, further research should enable forensic experts to more effectively investigate data from both device systems.

Our team's efforts to retrieve information via non-destructive methods resulted in limited information from the Amazon devices. Crucial previous user information was extracted however from the Google Home Mini device without knowledge of their username and password. This seems to be a major security flaw on the part of Google which can be exploited for further use in a forensics investigation.

References

1. April: Google support inquiry: case ID 9-6479000024453. [Online ChatSession]. https://www. google.com/. Accessed 25 Nov 2018
2. Choo, K.R., Goudbeek, A., Le-Khac, N.: A forensic investigation framework for smarthome environment. In: 17th IEEE International Conference On Trust, Security and Privacy In Computing and Communications/12th IEEE International Conference On Big Data Science and Engineering, pp. 1446–1451. https://doi.org/10.1109/trustcom/bigdatase.2018.00201 (2018)
3. Clinton, I., Cook, L., Banik, S.: A survey of various methods for analyzing the Amazon Echo. (Peer-reviewed Technical Report). The Citadel: Charleston, South Carolina. https://www. scribd.com/document/355663001/Clinton-Cook-Paper (2016)
4. Chung, H., Park, J., Lee, S.: Digital forensic approaches for Amazon Alexa ecosystem. Digit. Investig. **22**, S15–S25. https://www.sciencedirect.com/science/article/pii/ S1742287617301974 (2017)
5. Hendricks, D.: The history of smart homes. https://www.iotevolutionworld.com/m2m/articles/ 376816-history-smart-homes.htm. Accessed 22 Apr 2014
6. iFixit staff: Amazon Echo teardown. http://ifixit.com/, https://www.ifixit.com/Teardown/ Amazon+Echo+Teardown/33953 (2014)
7. iFixit staff: Google home teardown. http://ifixit.com/, https://www.ifixit.com/Teardown/ Google+Home+Teardown/72684 (2016)
8. Kim, E.: The inside story of how Amazon created Echo, the next billion-dollar business no one saw coming. https://www.businessinsider.com/the-inside-story-of-how-amazon-created-echo-2016-4. Accessed 2 Apr 2016
9. Koetsier, J.: Amazon Echo, Google home installed base hits 50 million; Apple has 6% market share, report says. Forbes https://www.forbes.com/sites/johnkoetsier/2018/08/02/amazon-echo-google-home-installed-base-hits-50-million-apple-has-6-market-share-report-says/# c3fd0c3769c2. Accessed 2 Aug 2018
10. Moore, P.: Homespeak. [Data file]. https://github.com/randomaccess3/homespeak (2018)
11. Moore, P.: Investigating rebel scum's Google home data. [PowerPoint Slides]. https://www. sans.org/summit-archives/file/summit-archive-1528488764.pdf (2018)
12. Orr, D., Sanchez, L.: Alexa, did you get that? Determining the evidentiary value of data stored by the Amazon Echo. Digit. Investiga. **24**, 72–78 (2018)
13. Spicer, D.: The ECHO IV home computer: 50 years later. http://www.computerhistory.org/ atchm/the-echo-iv-home-computer-50-years-later/. Accessed 31 May 2016
14. Stenhouse, E.: Smart speakers – How do they work? Evolving Science. https://www.evolving-science.com/intelligent-machines/smart-speakers-how-do-they-work-00483. Accessed 15 Nov 2017
15. Weinberger, M.: How Amazon's Echo went from a smart speaker to the center of your home. https://www.businessinsider.com/amazon-echo-and-alexa-history-from-speaker-to-smart-home-hub-2017-5. Accessed 23 May 2017

Digital Forensics of Raspberry Pi Used for MITM

Aman Gupta, Aishwarya Ulhas Desai and Ankita Sahai

Abstract Raspberry Pi, a credit card sized, low power consuming and cost-effective computer gained global recognition since its development in 2008. The capability to run GNU-based Linux distribution like Snappy Ubuntu, Kali, Raspbian and even Windows 10, has partly contributed to its popularity. However, these devices can also be exploited. In this chapter, we will discuss how a digital forensic investigator could investigate a case where a Raspberry Pi was used to conduct Man in the middle attack. Using an adversary model, we will demonstrate how one can perform a man-in-the-middle attack against users by creating a rogue wireless access point and capturing critical user information. We will try to analyze different operating system and application log files; files and data recovered using different digital forensic tools to deduce the timeline and what things have happened using recovered evidence. We will also see how different digital forensic analysis tools like Autopsy, FTK Access-Data and Bulk Extractor could be helpful in this case by performing benchmarking them against some key factors like recovered web history and cookies files, carved files, etc.

Keywords Raspberry Pi · Man in the middle attack · Rogue access point · Raspberry Pi forensics · Raspberry Pi log analysis

A. Gupta (✉) · A. U. Desai · A. Sahai
Department of Information Systems and Cyber Security,
University of Texas at San Antonio, San Antonio, TX, USA
e-mail: amannnng@gmail.com

A. U. Desai
e-mail: aishwaryaudesai17@gmail.com

A. Sahai
e-mail: ankita.sahai18@gmail.com

© Springer Nature Switzerland AG 2020
X. Zhang and K.-K. R. Choo (eds.), *Digital Forensic Education*,
Studies in Big Data 61, https://doi.org/10.1007/978-3-030-23547-5_8

1   Introduction

Raspberry Pi has gained global recognition since its development in 2008. One of the primary reason behind its because of its capability to run GNU-based Linux distribution like Snappy Ubuntu, Raspbian and even Windows 10. Since its inception various models have been introduced in the market with some of the latest models contains 64-bit ARMv8 Processor, wireless NIC cards, 1 GB LPDDR2 SDRAM and many more other features. These products have gained massive demand in the IoT market where these devices provide cost-effective and straightforward solutions to the problems. But these great capabilities also created new ways of exploitation of digital systems. There has been a lot of studies conducted which proved that these devices could be easily hacked or altered if they are not appropriately configured. There were many studies performed which show how Raspberry Pi could be used to perform an attack such as Man in the Middle attack [1, 2]. Now at the same time, it's getting difficult for a digital forensic investigator to figure out how to capture and analyze the data from Raspberry Pi.

In this paper, we are trying to contribute below things through the research we performed.

- Provide a way of a structural approach of investigating the Raspberry Pi evidence, where important logs files could be used to analyze important data points captured by Raspberry Pi before it has caught from a crime scene. To do so, we had set up our experiment in a way that we would create two different evidence images of Raspberry Pi which would have used various tools to perform Man in the middle attack.
- Perform a benchmark of forensics tools like Autopsy, FTK AccessData, and Bulk Extractor, and see how these tools work in showing the evidence image, what all files could be carved out and what web history files were still recoverable.

The paper is structured as follows. The next section talks about the existing literature on using Raspberry Pi for performing attacks, and cyber investigations of Raspberry Pi. The third section presents the research experimental setup and methodology. The fourth section presents findings for the each setup and different tools outcomes. The final section draws conclusions and outlines opportunities for future work.

2   Literature Review

Mortensen et al. [1] presented various types of attacks which could be performed using these miniaturized computers. According to Mortensen et al. [1], some of the attack vectors which could be facilitated by the miniaturized computers related explicitly to Raspberry Pi were:

Network Reconnaissance

Passive Traffic Capture: Attacker used this technique to capture the network packets using pcap and SSL stripper to sniff the sensitive data. Some of the main issues which the authors described while conducting this kind of attack are that it utilizes lots of power and storage.

Network Mapping: This type of attack takes advantage of tools like NMAP to survey the network. They found that it is easy to perform the network mapping through Raspberry Pi.

Active Attacks

DoS Attack: There are various ways to perform DoS attack based on the vulnerability present on the target machine. They tried to execute the DoS attack in which they routed all the traffic on /24 subnet to an invalid gateway. They were able to conduct a successful DoS attack until they had the power to run Pi.

MITM Attack: There could be various variants of MITM attacks are possible. The one which Mortensen et al. deployed was using "sslstrip" tool which was developed by moxie0.

Puangpronpitag and Sriwiboon [3] described simple and lightweight HTTPS enforcement to protect against SSL striping attack. HTTPS uses SSL/TLS to encrypt the data and protect it from sniffing attack. We can strip the SSL and then read the packet. It is done by using different tools like SSL stripping which acts as the man in the middle attack by intercepting the connection on one interface, stripping the SSL and then a connection to the other end using HTTP. In between, we can sniff the packet and read them as they are in plain text. The other attack is SSL sniff in which the server responds with the self-signed certificate to the client's certificate during SSL handshake. So, whenever the client uses the fake certificate while communicating information, the attacker can sniff and decrypt the SSL traffic because it has the private key to decrypt the encrypted traffic.

Agarwal et al. [4] studied advanced stealth man-in-the-middle attack in WPA2 encrypted Wi-Fi networks. They performed sophisticated MITM attack by sending ARP frame with GRPKY and high PACNUM value to the victim and poisons the victim's ARP cache. So, when victim sends data to the gateway, it sends data to the attacker and the attacker forwards data to gateway and sends a response to the victim. So as the victim gets a reply to its request, this method is transparent and hence more difficult to detect. It can breach broadcast multicast traffic protocol and sniff data.

Research by Vemi and Panchev [2] tried to implement a similar concept of performing a MITM attack using Unmanned Aerial Vehicle (UAV). They proposed a system in which Raspberry Pi was carried on a drone to set up rogue access and collecting credentials and data from the targeted wireless networks and connected devices.

WPS Pin Brute forcing Attack: Mortensen et al. [1] suggested and tried to exploit the vulnerabilities of WPS Pin implementation and tried brute-force attack, but due to power constraints they were not able to get any results.

Fully Automated Exploitation: Mortensen et al. [1] suggested to utilize Metasploit or custom set of procedures to exploit various know vulnerabilities of the targeted system. The success of these type of attacks was again depended on the power and resource availability to Raspberry Pi.

How wireless security works?

To move forward with our research, we also needed to understand how wireless security works. In the article by Vilius et al. [5], we were able to find all the wireless security protocols that exist. WEP is deprecated and can be easily cracked, but still many people still use it. WPA2 is the current standard which uses AES encryption and CCMP as the authentication protocol. Depending on the key length, it can be difficult to crack as compared to WEP encryption. From the study, we were able to find the approximate number of access points using different Wi-Fi security in an area. Looking at the data presented we see that WEP and WPA2 protocols are popular and many of the APs do not use any security mechanisms. Using this data, we can plan our attack accordingly.

Janbeglou et al. [6] described redirecting network traffic toward a fake DNS server on a LAN. They show us how to make DNS spoofing attack by spoofing MAC and IP address and then pinging continuously to switch. It will confuse the switch and make the switch send the data to ports of both authentic host and the attacker host. It happens due to lack of security mechanisms to check the trustworthy host. The authors of this paper conducted the attack on LAN. We can use this attack concept to redirect the user to the site that we want and compromise the credentials of the victim. The authors also demonstrate how to do session hijacking. Considering the user already has cookies and session ID, if the user goes to the same page again, the browser sends the cookie and Session ID to the web server to identify the users. During this process, if the attacker sniffs the Session ID and cookie, then the attacker can hijack the session or login as the authentic user.

Raspberry Pi as Honey Pot for collecting and monitoring threat vector

For our project, we sought out for more research about how the capabilities of a Raspberry Pi can be extended to deploy it to protect a network or a web application against potential malware. Djanali et al. [7] presented that Raspberry Pi can be used as a honeypot to increase the system's capability to detect and prevent attacks by trapping the attackers. In their experiments authors make a cluster of Raspberry Pi servers to improve its computational power as a cheap and effective honeypot. They developed a system to create web applications more secure from SQL injection attacks with the help of these honeypots. SQL injection is one of the most common security breaches in web applications. To carry out this attack, a hacker exploits the vulnerability between database and CGI layers by breaking into the website bypassing the proper authentication process to either steal valuable information or insert malicious data into the database. It poses a dangerous threat to various kinds of businesses, especially in the finance and e-commerce sector. Therefore, the authors of this study proposed to place Raspberry PI server honeypots as decoys to bait and trap the attacker while hiding the actual web server.

In another similar study Drake [8] has deployed a separate Raspberry Pi honeypot server to safeguard the network. In his paper, he explained how to use Pi-hole version 3 DNS server to protect the system from ad-based malware and how to sanitize a suspect's USB storage device to scan and remove harmful data it. He also outlines the steps on how the Raspberry Pi can be tweaked to perform as Kali Linux for wardriving to find security holes in the network. The act of driving around an area slowly searching for vulnerable Wi-Fi networks to exploit using any WiFi transceiver enabled device inside an automobile is called wardriving.

According to Drake [8] and another research paper by Balani et al. [9], an embedded Raspberry Pi security system operated by OpenCV can also be used to perform penetration testing on home routers and IoT devices to make sure they are safe from any exploitation. As we know that most of the routers and IoT devices support a fast and easy way to connect the wireless devices to the router known as WPS or Wi-Fi Protected Setup. Supposedly, WPS works only for wireless networks that use a password that is secured by encrypting with an eight-digit WPA Personal or WPA2 Personal security pins. This practice is a nightmare from the security point of view because the security pins for a standard WPS set-up can easily be can brute-forced by any hacker that has necessary tools and skillset. In these papers, the authors established that since it takes much longer time to execute Brute force and dictionary attacks on passwords on Raspberry Pi server than on a regular computer, it is advisable to use WPA2-AES encryption where ever possible and disable WPS on all the devices.

Raspberry Pi for performing MITM Wireless attack

Taking our discussion further let's review one of the MITM attacks which was performed using Raspberry Pi. Vemi and Panchev [2] experimented with Raspberry Pi as setting up it as a rogue access point (AP). They set up two wireless interfaces, one as a rogue access point and second as for monitoring and scanning available targets and their connected device. For setting first wireless interface as rogue AP, they used software tools like airmon-ng. For DHCP service an application named dhclient has been used, and 192.168.1.0/24 subnet is issued. They also changed the "key" setting and set to off which represents the password so that victim could get access to rogue AP. The second wireless interface is working in a monitoring mode to scan available targets which were implemented using "airmon-ng." A sub procedure of "airmon-ng" called as "airdump-ng" was used to monitor and capture packets. Once a victim got connected to rogue AP, he/she remains unaware of the change and using 4G network internet provided to Raspberry Pi, and the victim was still able to surf the web. Vemi and Panchev [2] were able to conclude that it is possible to hijack a WiFi session through Raspberry Pi mounted on a drone.

Digital Forensics of Raspberry Pi

Till now we discussed the aspect of how Raspberry Pi could be used as a threat agent. Work from Vemi and Panchev [2] supports the concept of use of Raspberry Pi as rogue AP. But now as digital forensics investigators, it's important to understand if we need to perform an investigation on Raspberry Pi then what all tools and resources are available for us. Unfortunately, until now little research has been conducted on Raspberry Pi forensics. Feng et al. [10], tried to perform extensive research on Raspberry Pi. As part of their investigation, they used Raspberry Pi 1, Model B revision 2. The first part of the paper tried to explain the six categories of the abstraction layer and its analysis.

Physical media layer: In the case of Raspberry Pi all the data could be found on SD card and analysis of this layer would deal with processing custom layout of data and as well recovering the deleted or overwritten data.

Media management layer: This deals with the logical partitioning of SD card storage and learning of arrangements of bytes and sectors.

File system Layer: In the case of Raspberry Pi it depends on the operating system file system that exists on SD card.

Application Layer: It involves forensic analysis of application data like config files, log files, user data files.

Network Layer: At this level of abstraction, data representation could be looked at physical or wireless medium.

Memory Layer: This layer translates the data at byte level for the system data.

Raspberry Pi has an installer manager caller New out of box software (NOOBS) which is used for the installation of operating system. Feng et al. [10] described the overall Raspberry Pi boot and operating system installation process which would be a helpful resource to understand the working of Raspberry Pi and understand the various partition created as part of the installation.

As part of the investigation some of the essential files and directories found by Feng et al. [10] were:

dmesg: Located at /var/log contains information about all the internal and external devices that were attached to the device during startup.

user.log: Located at /var/log provides information about various wireless access points the device has been connected.

kern.log: Located at /var/log contains information about kernel operations, and if SD card storage is only investigated then it helps to know the model of Raspberry Pi device to which it is connected.

history.log: Located at /var/log/apt contains information about the installing of installed, uninstalled, updated packages on the devices.

.bash_history: Located at /home/pi includes the history of all the commands executed at the bash command prompt.

shadow: Located at etc/shadow contains users login credentials.

One of the observations they found was that Raspberry Pi does not give much of operating system footprint details while performing network scanning through tools like NMAP as compared to desktop or PCs. Some of the observations made for the forensics toolkit on SD card installed with Raspbian OS were:

ProDiscover: It was only able to view the FAT partition of the SD card and hence could not be used to perform data acquisition.

FTK: It does not copy the master partition table of a physical media which contains start and end of sectors of all partitions exist in media as well master boot record. FTK only shows NTFS and FAT partition and hence unable to show MPT.

Encase: This tool was able to create an image of the entire SD card but unable to access the extended partition for analysis.

3 Experimental Setup and Tools Used

To set up our experiment, we choose to select Kali as our base Operating system as it has excellent support for different tools used to exploit the vulnerability or create rogue access points. Kali has a specific operating system trimmed to run on Raspberry Pi. We used Kali Raspberry Pi-2018.3-RPI3-NextMon version for our setup. After picking the operating system, our next task was to select tools which could be used to develop a rogue access point and perform Man in the Middle Attack.

A. *Man in the Middle Attack details*

In this experiment, Raspberry Pi was used to perform Man in the middle attack by creating a rogue access point and obtaining sensitive information. Figure 1 shows the experimental Raspberry Pi used. Power to the device was given through an AC to DC adapter. Internet connection was provided over Wi-Fi by connecting Raspberry device to an access point created by us. We also connected an external wireless adapter in case the internal wireless card didn't work.

We used Samsung Galaxy S8 as our test target device and used Firefox focus V_7.0.13 as the browser through which we surfed through different websites. The websites we planned to browse through were Facebook, Discover credit card and Netflix.

While setting up the access point we provided a common and known SSID name to which our test target device was connected earlier. Two wireless ports Wlan0 and Wlan1 of Raspberry Pi device were set up in such a way that wlan0 would act as a rogue access point and wlan1 would be connected to the real internet. Wireless port setup was done through AP module in case of FruityWifi and hostapd for the Wifi Pumpkin. The step by step process to achieve this were as follows:

- The user gets automatically connected to the rogue access point.
- Using his browser (which was Firefox focus V_7.0.13 in our case) started connecting to websites which were HTTP.
- The request is logged in the device through sslstrip/sslstrip2 module and all the details were logged into the application logs.
- Then the request is made secure and forwarded to the real internet through wlan1 port to which internet connection was provided. Doing so the server thinks that the connection was secured.

Fig. 1 Raspberry Pi 3 Model B+ experimental setup

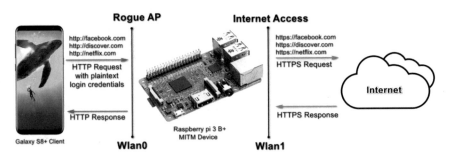

Fig. 2 Pictorial representation of attacks

- After a considerable amount of time surfing through different websites, we stopped the Raspberry Pi (Fig. 2).

(1) *Man in the Middle Attack using FruityWiFi*:

We have taken the Samsung 16 GB SD memory card and formatted using default format options of Windows. Then we installed Kali for Raspberry Pi on it. For the first experimental setup, we selected FruityWiFi which is an open source tool to audit wireless networks. It could be used to deploy advanced attacks by directly using the web interface or by sending messages to it. At first, it was created to be used with the Raspberry-Pi, but it can be installed on any Debian based system. We used version 2.4 for the experimental setup. FruityWifFi has numerous inbuilt modules to

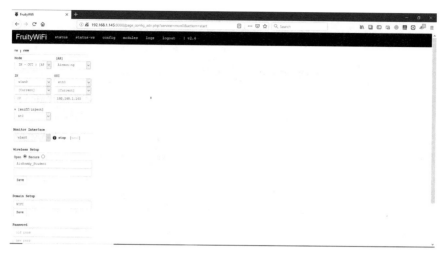

Fig. 3 FruityWiFi access point setup

perform wireless network audit and conduct the wireless attacks. The module which we planned to use for this experiments were:

AP: Acronym of Access Point, is a module used to create a rogue access point to navigate all the traffic through Raspberry Pi instead of authentic access point. It has two nodes, one of them is connected to real internet and second act as a rogue access point.

sslstrip: This module is used to perform a man in the middle attack where Raspberry Pi becomes a proxy between the user of rogue AP and the webpage user is visiting. It strips off the SSL configuration present on the website and makes an https website into an HTTP website, establishing all the traffic communication in plain text and log the request and response details. FruityWiFi's sslstrip module includes the options Inject and Tamperer, but as part of our experimental setup, we are not using it. We also updated the Kali config file so that FruityWiFi is even started when the system boots up (Fig. 3).

(2) *Man in the Middle Attack using WiFi Pumpkin*:

Similar to the first setup, in this experiment also we used Samsung 16 GB micro SD card on which we installed Kali Linux. The SD card was wiped before use. We used Wi-Fi Pumpkin tool which makes Man in the middle attacks easy to execute. Raspberry Pi works as a MITM device between the target and the internet.

SSLStrip2: This module is the same as Sslstrip, but it works better in scenarios like when the communication is intercepted when the TLS connection was already initiated. When used with DNS2proxy we can implement HSTS attack. In this attack, we strip the Strict-Transport-Security header field (Fig. 4).

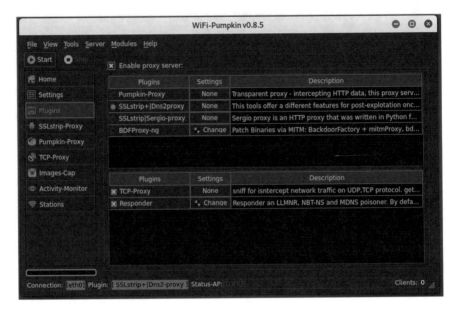

Fig. 4 Pumpkin WiFi sslstrip configuration

DNS2Proxy: This tool is used to redirect the DNS request to the fake DNS server, and then the attacker intercepts the request and send it to real DNS server to the real name and sends the response back to the fake domain name and real IP back to the user. When the browser receives the DNS response, it searches if the domain is accessed by HTTPS by checking HSTS list of domains. Because the user has a fake domain, it makes an HTTP connection with real IP address (Fig. 5).

B. *Tools Used*

In this section, we are going to discuss in brief about the tools we are going to use as part of collecting and analysing the evidence image.

(1) *Autopsy 4.9.1*:

The autopsy was one of the three tools selected for analysing the image files. It is an open source tool which is easy to use. It is a GUI-based program which allows analysing the hard drives, disk images, and smartphones efficiently. For the examination of the evidence, Autopsy 4.9.1 has been used. Autopsy provides various custom views apart from viewing the file system structure. Some of the views which could be useful from investigation perspective are:

- EXIF Metadata (EXIF, i.e. Exchangeable Image File Format)
- Encryption Detected
- Encryption Suspected
- Extension Mismatch Detected
- Web Bookmarks

Fig. 5 Pumpkin WiFi access point configuration

- Web Cookies
- Web History
- Web Search

(2) *Bulk Extractor 1.5.0*:

It is an image analysis tool which can be used to identify the traces of the incident by collecting all the data like domains, emails. It needs an image of the disk which can be further be used to analyze the event.

This tool can find email addresses, telephone numbers, passwords, as keys, domain information and addresses, URL searches and lot more. It gives a good knowledge about from where the incident started by listing the history of the activities and some essential data to the analysis.

(3) *FTK AccessData 6.0*:

AccessData FTK scalable and analytical digital forensic investigation platform. It is well known for its various capabilities like perceptive interface, stability, email data analytics, and user views that can be tailored according to the user's requirements. FTK also provides facilities to perform malware analysis and neoteric data visualization. The increased prevalence of inaccuracy and unreliability in digital data supports the need for scientific validation and verification of digital forensics evidence within the legal system.

Table 1 Hardware and software version details

#	Tools name	Version
1	Raspberry Pi 3	Model B+
2	Kali Raspberry Pi	2018.3-RPI3-NextMon
3	Fruitywifi	2.4
4	WiFi Pumpkin	0.8.5
5	Autopsy	4.9.1
6	FTK Imager	4.2.0.13
7	FTK AccessData	6.0
8	Bulk Extractor	1.5.0

The characteristics of digital forensics investigation are such that the analysis of evidence largely depends on the data available, and the methodology and tools used. The forensic Tool Kit is fortified with features that assist the investigators in tracing specific type of data based on its distinct characteristics. Some of these features include file carving and password recovery capabilities.

(4) *FTK Imager 4.2.0.13*:

FTK Imager is a forensic image creator build by AccessData. This tool can be used to create the forensic image of any physical or logical disk which can be further be used to examine in different analysers like FTK toolkit, Bulk Extractor and Autopsy. There are features like capturing the volatile memory of the device like capturing the data in RAM and other volatile data like running processes and network connections. It calculates MD5 hash which can be used to verify if the image is changed in future. We can view the file in read-only mode so that we can analyze the image without changing its contents and we can even export files if needed.

Table 1 presents the details of the software and hardware we used for this experiment.

4   Forensics Analysis

In this section, we would be discussing the analysis of the data extracted and findings for the investigation. For our experiment, we used McKemmish [11] approach for conducting the research. As per McKemmish [11], there were four phases of the forensic investigation.

- Identification
- Preservation
- Analysis
- Presentation

A. *Identification and Preservation*

As our problem statement says, in the crime scene Raspberry Pi was present. This device has been used to perform Man in the middle attack by creating a rogue access point and obtaining sensitive information. To perform the digital forensics on Raspberry Pi which was collected from a crime scene our job as a forensics investigator was to identify the evidence. In our case Raspberry Pi gathered was Raspberry Pi 3 Model 3+ and contains Samsung 16 GB class 10 SD memory card.

After collecting the evidence from the crime scene, the evidence image of the SD card, present in the Raspberry Pi must be created. Acquiring the disk image has been referred to as the preservation phase for our experiment. We used FTK Imager to create the image of the SD card as it is considered as the fastest and most reliable imaging tool by Shaver's [10]. FTK imager 4.2.0.13 has been used to create the image of the SD card.

We would also like to justify the rationalizations for not performing live analysis on Raspberry Pi. The reasons for not analysing the live memory of Raspberry Pi were:

- No or limited resources for performing live analysis on Raspberry Pi.
- Analysing the memory of Raspberry Pi when it is live is risky because the power supply could be interrupted.
- It could be possible that raspberry Pi was connected and monitored remotely. In that case, analysing and performing the investigation live could create a risk that the criminal wiping the device remotely.

Considering all these points, it is better suited to create the disk image file offline. This will preserve the evidence in two ways:

- First the offline image creation was done through FTK images which uses MD5 hashing to preserve the integrity of the image.
- Second, performing offline investigation will reduce the chances of contaminating the evidence by the user who could login remotely.

B. *Analysis*

In this section, we are going to discuss about the analysis we performed. In this experiment, we used two approaches to analyze the evidence images. The first approach we took was the analysis of various log files created by the operating system and applications. Through this, we aim to build the timeline of the multiple events occurred in the evidence image. We would also try to correlate various log files for analysis and try to narrow down the intent of the user of the device.

The second approach we took was to analyze the evidence image across various forensics tools discussed in the above section. The intent of doing so is to come up with a benchmark of different tools for and see how they behaved when they were used to analyze the evidence image captured from Raspberry Pi.

Table	Thumbnail					
Name	ID	Starting Sector	Length in Sectors	Description	Flags
vol1 (Unallocated: 0-0)	1	0	1	Unallocated	Unallocated
vol2 (Win95 FAT32 (0x0c): 1-125000)	2	1	125000	Win95 FAT32 (0x0c)	Allocated
vol3 (Linux (0x83): 125001-30308863)	3	125001	30183863	Linux (0x83)	Allocated
vol4 (Unallocated: 30308864-31291391)	4	30308864	982528	Unallocated	Unallocated

Fig. 6 Evidence 1 partition details

(1) *Analysis of Evidence image through log files*

For this part, we utilized Autopsy tool to fetch the log files as this tool was able to get the files in the same way as it was present in the operating system.

(a) *Evidence 1's Analysis*:

To start with, we first needed to figure out which operating system and file system has been installed in the evidence image.

- On analysing the image in the Autopsy tool, we figured out that Operating system was identified as Linux. This could be found by looking at the allocated volumes information (Fig. 6).
- */etc/issue* and */etc/issue.net* contains the information of the operating system. By analyzing it we found the version to be Kali GNU/Linux Rolling.

After this, our next steps were to find out what has happened and how.

- **history.log** file contains the information about installed, uninstalled and updated packages on the devices. We started analyzing the history.log which could be found at */var/log/apt/* and found some of the key information analyzing the file:

  - At 2018-10-03 05:21:38(CST), the FruityWiFi application was installed in the evidence image by running the command *apt-get install fruitywifi*.
  - At 2018-11-14 16:20:53(CST), the xrdp application was installed by running the command *apt-get install xrdp*.

- **auth.log**: While analyzing the image we could not see any user profiles, which means that only root user account has been used to assess the operating system. We verified this by analyzing the auth.log file present at */var/log/* and checking for any deleted user. We do not see any instance of deleting a user.

As we know now that there is only one user which was used to access the system, our next step was to find the password for the user and check if any commands were run by him using command line interface (CLI).

- **shadow and passwd file**: We could find these files at the /etc/location. These files contain the credentials information for all the user. All this information is hashed and then saved to the file. To see the credentials, we used Johnny the ripper tool to extract the details from both the records. We were able to look at the credential

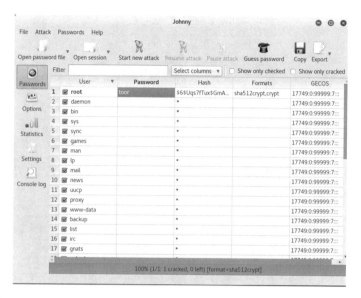

Fig. 7 Cracking the passwd and shadow file

details and all the username. This information could be used to access the Operating system and perform live forensics which is currently not part of our scope. Below image shows the result of the tool after cracking the passwords (Fig. 7).

- **bash_history.log**: This file contains the history of all the commands executed at the bash command prompt. This file could be found at the user profile location. As we have already figured out that for this case, the evidence image shows that all the actions were performed using root credentials, we found the bash_history.log file at */root/*. Some of the vital information we found by analysing the file was:

Root user account was used to install fruitywifi, and some command was executed to make FruityWiFi run automatically after the start of the system. *update-rc.d* command enables the it to run the FruityWiFi after system boots up.

A few more commands were executed, which makes it look like the user was trying to access the sslstrip.log file, which is one of the modules present in the FruityWiFi application. The user navigated to */var/log/FruityWiFi* location to view the log files. These were one of the few last commands fired from that profile as after that we found system was rebooted, and no other actions were recorded (Fig. 8).

From the above findings, we got a clue that the root user is trying to access and check the sslstrip.log files. Sslstrip is one of the modules present in the FruityWiFi where SSLStrip is used to perform a type of MITM attack that forces a victim's browser into communicating with an adversary in plain-text over HTTP, and the adversary proxies the modified content from an HTTPS server. We started digging deeper to check what else we could get from those files. We started analysing the log files generated by fruitywifi. We understood that FruityWiFi generated the real-time log file and kept that file at *the /var/log/FruityWiFi* location.

```
vi ngrep.log
vi sslstrip.log
vi ngrep.log
vi sslstrip.log
cd /var/log/fruitywifi
ls
less sslstrip.log
tail-f sslstrip.log
tail -f sslstrip.log
less sslstrip.log
tail -f sslstrip.log
reboot now
```

Fig. 8 bash_history.log file content

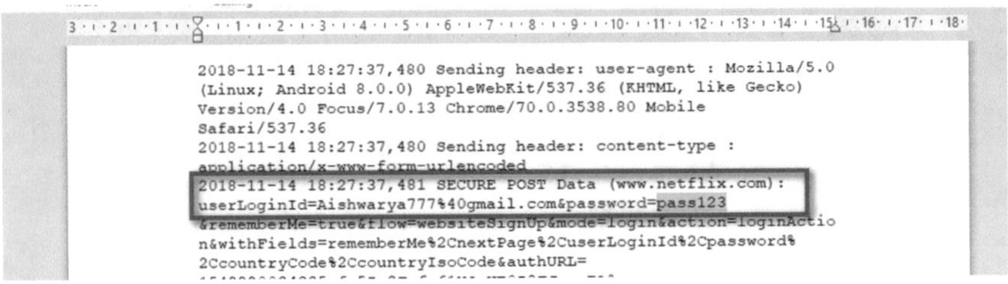

Fig. 9 Logs file suggests that this tool can capture the sensitive information

Whereas FruityWiFi archives each log file for each module at below location:

$$/usr/share/fruitywifi/www/modules/ << module$$
$$name >> /includes/logs$$

Where *<<module name>>* is the module we need to check for. As the bash_history.log suggests that user was trying to access the sslstrip module logs, we started concentrating on that, which could be found at location */usr/share/fruitywifi/www/modules/sslstrip/includes/*logs. While analysing those logs, we could find that some of the credentials were captured in the log files which could help in understanding the motive of the user. It suggests that user wants to obtain the credentials of the users gets connected to the rogue access point (Fig. 9).

- **xrdp.log**: As we saw in history.log that the xrdp application was installed in the captured evidence, we started analysing the xrdp.log file. xrdp.log is generated by XRDP application which is used for remotely connecting a Linux machine. This log could be found at /var/log/xrdp.log location. After analysing the logs, we found the entries whereas a part of setting up the experiment we logged into the machine.

  – A login attempts at 20181114-16:25:35 by AMAN PC.
  – A login attempts at 20181114-17:03:28 by ANOMEE (Fig. 10).

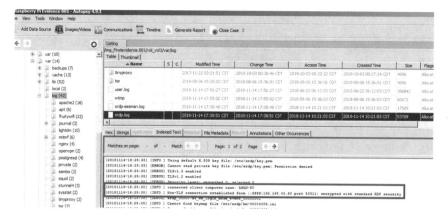

Fig. 10 Login attempts through XRDP

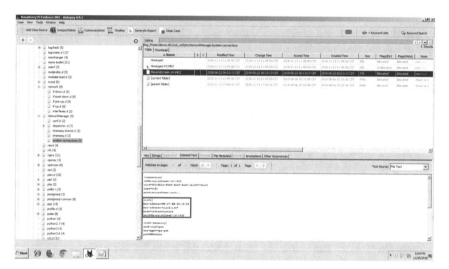

Fig. 11 Files present in etc/NetworkManager/system-connections/

Above details could be admissible evidence in the court of justice to prove that the suspect is guilty.

- **etc/NetworkManager/system-connections/**: This is an important location where we could find the details of the wireless access points to which the Raspberry Pi was connected. While analysing the log file, we found that it has been linked to 2 access points, and from that, it could help in narrow down the suspects (Fig. 11).

Fig. 12 File system and operating system details

(b) *Evidence 2's analysis*:

Like evidence one we need to figure out the operating systems and file system.

- On analyzing the image in Autopsy tool, we figured out that the it is a Linux Operating system (Fig. 12).
- By doing further analysis of *etc/issues* and *etc/issues.net* we found the version to be Kali GNU/Linux Rolling.

After this, we started analyzing history and auth logs to track the trails left by the user of the collected evidence.

- **history.log**: From the history.log we found nothing interesting after analyzing.
- **auth.log**: We wanted to check this log file to check if there are any other user than root. While investigating the logs, we found that on Nov 20 17:32:24 a ghost user has been added.
- **shadow and passwd file**: To see the credentials we used Johnny the ripper tool to extract the details from both the records. We were able to look at the credential details and all the usernames. This information could be used to access the Operating system and perform live forensics which is currently not part of our scope. Below image shows the result of the tool after cracking the passwords (Fig. 13).
- **bash_history.log**: From auth.log we found that there is a new user added so that we would analyze the bash_history.log file for root and ghost user.
  Root user: After analysing the bash_history.log file for the root user we found that:

  – Root user added the ghost user.
  – Installed WiFi pumpkin module on the system.
  – Then accessed the credentials.log file (Fig. 14).

  ghost user: We could not find any log file which means that there was no activity performed by the ghost user.

From the above details, we could understand that WiFi pumpkin module has been installed in the system. It is also evident that there was no action performed by the ghost user using the command line as there was no bash_history.log file present for the same. Again, digging deeper, we could see that user accessed the logs for WiFi Pumpkin. Taking the lead from that, we found two locations where we could find logs generated by the WiFi pumpkin.

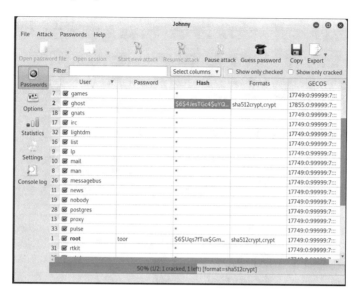

Fig. 13   Cracking the passwd and shadow file for evidence 002

```
new 1    Instr.txt    new 2    new 3    new 4    new 5    .bash_history    auth.log
1   adduser --home /ghost ghost
2   git clone https://github.com/P0cL4bs/WiFi-Pumpkin.git
3   ls
4   cd WiFi-Pumpkin/
5   chmod +x installer.sh
6   sudo ./installer.sh --install
7   sudo wifi-pumpkin
8   sudo apt-get update
9   shitdown now
10  shutdown now
11  sudo wifi-pumpkin
12  apt-get update
13  apt-get install mitmproxy
14  ifon
15  ifconfig
16  clear
17  ifconfig
18  ls
19  cd WiFi-Pumpkin/logs/AccessPoint/
20  ls
21  vim credentials.log
22  cd WiFi-Pumpkin/
23  sudo wifi-pumpkin
24  |
```

Fig. 14   bash_history.log file for evidence 002

```
GET https://104.72.133.23/images/header-nav-footer/bg-tab-content-shadow-wrap.png
{'content-length': '0', 'accept-language': 'en-US', 'referer': 'http://www.discover.com/css/

GET https://23.207.55.244/customersvcs/universalLogin/ac_main?ICMPGN=HDR_LOGN_CC_LOGN
{'content-length': '0', 'accept-language': 'en-US', 'referer': 'http://www.discover.com/m/',

POST https://23.207.55.244/customersvcs/universalLogin/signin
{'origin': 'http://webportal.discover.com', 'content-length': '137', 'accept-language': 'en-
userID=Ashleygupta&password=yoyohoneysingh&accountType=&userTypeCode=C&pm_fp=&currentFormId=
GET https://23.207.55.244/navl/notification.html
```

Fig. 15 Credentials retrieved from the logs of the proxy server

Fig. 16 Files present in etc/NetworkManager/system-connections/for evidence 002

$$usr/share/WiFi\text{-}Pumpkin/logs/$$
$$usr/share/WiFi\text{-}Pumpkin/$$

Looking deeper into the log files we found sensitive information captured by the tool (Fig. 15).

- **etc/NetworkManager/system-connections/**: As evident from evidence one, we found the details of the wireless access points to which the Raspberry Pi was connected. While analyzing the log file we found that it has been connected to 1 access point this time, and from that, it could help in narrow down the suspects (Fig. 16).

(2) *Analysis of Evidence image through tools*

For this part, we utilized three tools to analyze each evidence image. The tools we used are Autopsy, AccessData FTK, and Bulk Extractor.

(a) *Autopsy*

Evidence 1
Autopsy classifies the extracted content in various categories.

EXIF Metadata: This module of Autopsy provides information for images that can contain geolocation data for the picture, time, date, camera model and settings (exposure values, resolution) and other information.

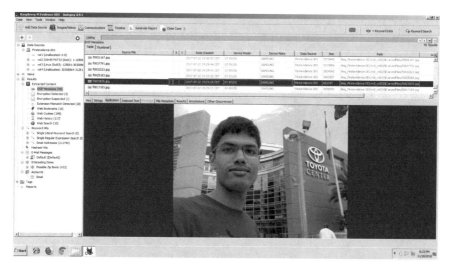

Fig. 17 EXIF Metadata content extracted by Autopsy

While analysing the data recovered by the Autopsy tool, we could find images which were not part of the operating system but taken earlier than the operating system was installed on SD card. We could see the image metadata like date created, device model and device makers. The number of carved images by this evidence image was 45 (Fig. 17).

Web Bookmarks: After analyzing bookmark section we could see a list of 16 bookmarks recovered by the tool, with the details like URL, Title, date created and program name (Fig. 18).

Web Cookies: Analyzing this section we could see that web cookies recovered from the evidence image.

Web History: Analyzing this section helps us to understand the browsing pattern of the suspect and what websites or URLs have been opened through the browser.

Analyzing the web history and pattern of Google searches performed, we could say that suspect was trying to know details about installing FruityWiFi in the Raspberry Pi and running different modules of the same (Fig. 19).

Web Search: Analyzing the Web search results recovered from the images reiterated the finding we had in the previous section that the suspect was trying the search on how to install FruityWiFi on Raspberry Pi.

Evidence 2

For evidence 2 we were predicting that fewer data would be recovered by the tool as this time the SD card was formatted by Wiping tool and not formatted by the default Windows tool. Some of the critical data extracted by the tool were:

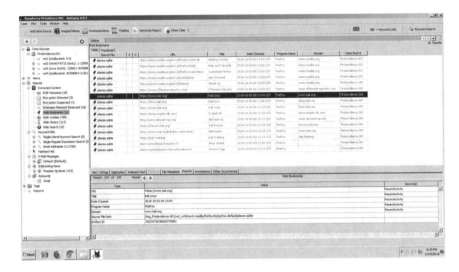

Fig. 18 Web bookmarks content extracted by Autopsy

Fig. 19 Web history extracted from evidence 1 by Autopsy

EXIF Metadata: This time we could not find any images which could be used as evidence. The reason could be because the SD card has been wiped out.

Web History: The tool was able to extract the web history data exceptionally, and we could see the browsing pattern of the user. Analyzing the browsing pattern, we could say that the user was googling about how to install WiFi pumpkin on Raspberry Pi (Fig. 20).

Fig. 20 Web history extracted from evidence 2 by Autopsy

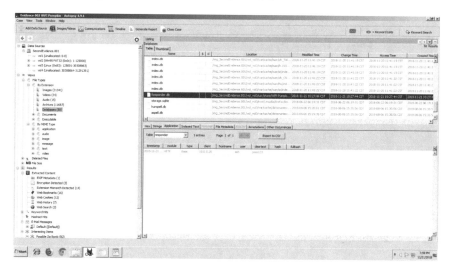

Fig. 21 Database extraction using Autopsy tool

This section of the tool presents us the web searches performed using the browser in the Operating system. The tool was able to extract the web search results, and the results were like what we got from web history.

Database: We also analyzed database extracted by the tool, and while examining the database we came across the Responder database created by the responder module of WiFi Pumpkin. From that, we obtained the username and password of that module stored in the database (Fig. 21).

(b) *FTK AccessData*

Evidence 1

We opened a new case for our analysis of the evidence file with AccessData FTK 6.0. Once we create a new case, we were presented with a prompt to enter the case name. Here, we can enter the reference to this case if needed and this tool also has the capability to add interspecific information about the case in the description field once this has been completed, we got a couple of additional options that we could select like the detailed options that's where we can refine our evidence processing

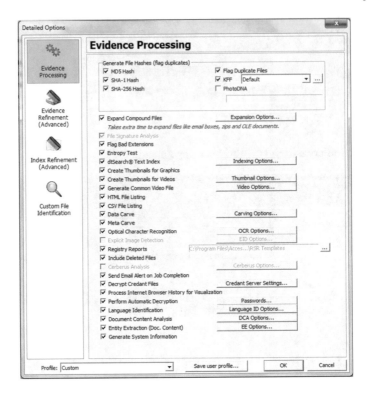

Fig. 22 Setting up FTK AccessData

criteria. Many of the options given can be toggled on and off based on the refinement of this bit of a case (Fig. 22).

Since our investigation was not time sensitive, we chose various options in this phase for processing of the case. If we had a constraint like that, we could have left out the time-consuming tasks to perform additional analysis later. One of the options that we chose during case creation was Data Carving. Data carving is the process of deriving deleted or altered files from the file structure. The deleted files can be partially or recreated by identifying the file header and footer information in the unallocated clusters.

Performing data carving on the evidence file takes a considerable amount of time. Since, we had good reason to believe that the perpetrator might have intentionally deleted some data to cover his tracks. Suspecting that we could gain vital information from this, we opted to check the data carving option (Fig. 23).

It took almost 45 min for the FTK to process the evidence image. We also made sure to maintain the data integrity of the acquired image by using the Secure Hash (SHA1) and Message Digest hash (MD5) algorithms.

It can be seen from the image below (Fig. 21) that the file structure consists of different levels, and the data in each level is carved from the respective location using

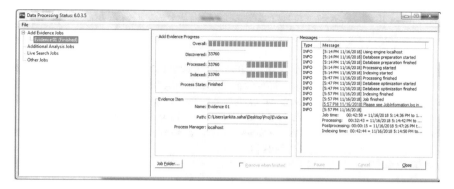

Fig. 23 Status of data processing while adding evidence 1 to the case

Fig. 24 Overview tab giving a breakdown

the Forensic Tool Kit (FTK) carving function. One significant difference that can be observed is that the tool shows the disk structure which is somewhat different from that of any hard disk drive. Another notable thing here is that the data craved from the SD card might be incorrect or it might have been carved out from the wrong location [12].

One forensically interesting aspect of investigating digital data through FTK is to find the intentionally dark or hidden data as it exists on modern file systems. Unfortunately, it works best for Windows and UNIX file systems, so we could not find anything of interest in our analysis. However, one of the remarkable traits that we see signs from the investigation standpoint was that the tool was able to carve out the data using file signatures (Fig. 24).

The results of the analysis are summarized and presented to the investigator in different categories on the Overview tab. This data can then be utilized to analyze, validate, and report on the artifacts that point towards Bowser and folder activities. System registry changes can also be examined to establish a timeline about the suspect, i.e., what all did he do, and how and when he did it. For instance, we can see that one of the carved files is an Excel spreadsheet. Even though the data in the file is not of much importance to our investigation, it still reflects the potential of FTK as an investigative tool.

Because of the limited ability of the tool to analyze the file system and the internet browsing history, we could not determine the accuracy of the artifacts from the FTK utility alone. We reckon that if we combine other complimentary forensic tools to extract the data and analyze it. One such example could be configuring SQLite database to deconstruct the data relevant to the case from the files pulled out from the images and web cache artifacts [13].

Evidence 2

Just like Evidence 1, we selected all the option while preparing the case for Evidence 2. However, this time it took a little longer to finish the process. FTK was able to obtain the Email attachments, graphics, internet chat files, documents, for Evidence 2 as well. Similar to Evidence 1, the information extracted from this image is also inconclusive for our experiment. However, there were slight differences in the results of the experiments, like the images found in this case were mostly thumbnails from the applications installed on the operating system, instead of the actual gif, jpeg, files as seen on the previous evidence. It could be because this time around the suspect might have used the 'Overwrite format' while formatting the SD card.

Additionally, FTK does have an interesting aspect to it that it can also categorize the files based on the content of the document. It can bring forth useful information about the case. While this may not be important to our case itself, but it gives us some additional insight as to how some of the intrinsic case details may prove useful when looking for evidence. The recovered content of these might be used to aid in substantiating the evidence. This content includes the user-related information such as social security numbers, e-mail addresses, phone numbers, credit card information or other personal information that is found by parsing through the recovered files (Figs. 25 and 26).

Registry Viewer is another remarkable tool from Access Data that can be employed to uncover the passwords for various user accounts and to access other protected Reg-

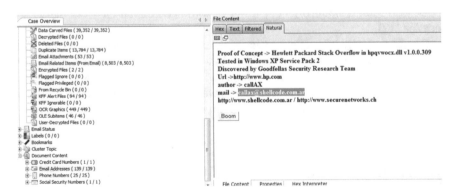

Fig. 25 Email address extracted using the AccessData tool

Fig. 26 Credit card info extracted using the AccessData tool

```
..C....................MESSAGE=wlan0: Trying to associate with
34:8f:27:68:d6:48 (SSID='MaverickCreek-14-1402' freq=2412 MHz)..
........k.......B.._¢@............C.......C................
_SOURCE_REALTIME_TIMESTAMP=1529665996649305...................
```

Fig. 27 Client association attempt

istry files associated with the password recovery process. A look into the registry with Registry Viewer might also show what all programs were installed and uninstalled from the system.

Similar observations from both Evidence 1 and 2 confirm that it is possible to extract meaningful digital information/metadata through FTK. However, it was only partially successful in building a weighty history of events that could be used in construing the digital artifacts obtained from Raspberry Pi that could support the investigation. Nevertheless, other related studies [13] have also concluded that if FTK was used in conjunction with other tools such as SQLite database browser, VMware player, Access Data's Registry Viewer, it could be much more useful to try and recover some other data of interest.

(c) *Bulk Extractor*

Evidence 1

Network Logs: We were able to find networking logs showing the information about SSID of the APs, MAC address of the devices and the AP's, all the probe requests send by the client devices. We were able to see the clients attempting to connect to the SSIDs on 2.4 MHz frequency along with the timestamps, so we were able to tell the specific time when the client was trying to connect to AP. Below is the example (Fig. 27).

Images Carved and PII data extraction: This tool carved 13 images from the forensic image which is very little as compared to other tools. When we see the PII data extracted, the tool was able to find bitcoin address which can be used to find the traces of the attacker. However, in this case, the Bitcoin IDs belonged to the authors of the open source tools used by an attacker (Fig. 28).

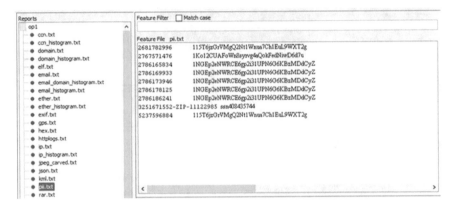

Fig. 28 PII data extracted by Bulk Extractor for evidence 1

```
an0 with hwaddr bc:f6:85:03:36:5b and ssid "Rowdy
an0 with hwaddr bc:f6:85:03:36:5b and ssid "Rowdy
an0 with hwaddr bc:f6:85:03:36:5b and ssid "Rowdy
id=PumpAP\x0Abssid=BC:F6:85:03:36:5B\x0AinterfaceAP=wla
_Student1\x0Abssid=BC:F6:85:03:36:5B\x0AinterfaceAP=wla
_Student1\x0Abssid=BC:F6:85:03:36:5B\x0AinterfaceAP=wla
[MjExMzA=] STA dc:ef:ca:73:64:0d IEEE 802.11: as
P-STA-CONNECTED dc:ef:ca:73:64:0d\x0Awlan0: STA dc:e
4:0d\x0Awlan0: STA dc:ef:ca:73:64:0d RADIUS: startin
67 : wlan0: STA dc:ef:ca:73:64:0d IEEE 802.11: as
```

Fig. 29 Client connecting to Rouge AP

Evidence 2

Network Logs: Like the analysis of evidence 1, we can see that the client connects to Rowdy_Student1 AP and the tool was able to find the trails of the event (Fig. 29). **Images Carved and PII data extraction**: This tool carved 15 images from the forensic image which is very little as compared to other tools. When we see the PII data extracted, the tool was not able to find any PII information on the second forensic image (Fig. 30).

5 Conclusions

There are three major conclusions that can be drawn from these experiments. Our analysis supports the initial hypothesis that it is possible to use Raspberry Pi as an instrument in carrying out Man in the Middle attacks and digital forensics can be used to extract admissible methods for evidence collection and analysis.

The experiments rendered that it is possible to retrieve or extract significant digital information in form of system logs, metadata and PII from evidence images of

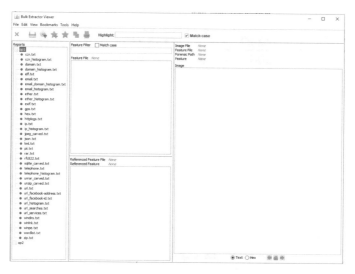

Fig. 30 PII data extracted by bulk extractor for evidence 2

Raspberry Pi. These significant results obtained from the investigative tools could be used as an admissible method for evidence collection in similar other situations.

The forensic prerequisites, techniques, and test affirmation used to create forensic images of the memory card and then used 3 different tools to analyze the extracted artifacts enabled the identification, collection, and analysis of digital evidence for our study. This digital evidence was in the form of log files which could help a forensic investigator to build a significant timeline of events that led to the incident and could possibly incriminate the suspect.

Other valuable information gathered from the study was that we were able to benchmark the three tools that we used to analyze the data based on how efficient the tools were in processing the image and present the correct metadata values which was indicative of the details about the activity that has occurred.

A comparison of the tools we used to analyze the digital artifacts extracted from the evidence images was done to evaluate their performance against certain parameters. Through this study we found that Autopsy is good option in our case as it is open-source, can generate reliable analysis report and has better flexibility to do manual analysis. Therefore, this tool might help in acquiring relevant digital evidence from Raspberry Pi in the future. Some interesting findings from the evaluation is shown in Table 2.

For future work, this study can be extended further to perform live memory analysis of Raspberry Pi so that we can collect the running processes in RAM. We can all network connections made by Raspberry PI and see how they will affect digital forensics investigations. This will enable us to determine that the attacker wants to exploit. It will aid in reducing the investigator's time and efforts in figuring out the

Table 2 Comparision of forensics tools used

Parameter/tool	Autopsy	FTK toolkit	Bulk extractor
File system structure	Able to extract the file system structure	Not able to create the file system structure	Not able to create the file system structure
Networking logs	Did not created special networking logs. Navigated through the file system to find the required evidence logs	Did not create special networking logs	Extracted to Ethernet logs
Web artifacts	Created different sections for Web cookies, histories, and searches performed in the browser of Raspberry Pi. Looking the entries it was easy to understand that the browser used was Firefox	Did not find any Web artifacts	Extracted session IDs and cookies. It was presented as logs written in text file
Carved out deleted images	Only 45 deleted images were carved out	More than 30,000 deleted images were carved out	None
Processing speed of evidence image	Average time to process the image was around 15 min	Average time to process the image was around 45 min	Average time to process the image was around 60 min

particulars of the case he is working on. We can also broaden the scope of bench-marking process by including more forensic tools. These two added features can be the center of the next part of this research.

References

1. Mortensen, C., Winkelmaier, R., Zheng, J.: Exploring attack vectors facilitated by miniaturized computers. In: Proceedings of the 6th International Conference on Security of Information and Networks, pp. 203–209. ACM (2013)
2. Vemi, S.G., Panchev, C.: Vulnerability testing of wireless access points using unmanned aerial vehicles (uav). In: European Conference on e-Learning (2015)
3. Puangpronpitag, S., Sriwiboon, N.: Simple and lightweight HTTPS enforcement to protect against SSL striping attack. In: 2012 Fourth International Conference on Computational Intelligence, Communication Systems and Networks (CICSyN), pp. 229–234. IEEE (2012)
4. Agarwal, M., Biswas, S., Nandi, S.: Advanced stealth man-in-the-middle attack in WPA2 encrypted wi-fi networks. IEEE Commun. Lett. **19**(4), 581–584 (2015)

5. Vilius, K., Liu, L., Panneerselvam, J., Stimpson, T.: A critical analysis of the efficiencies of emerging wireless security standards against network attacks (2015)
6. Janbeglou, M., Zamani, M., Ibrahim, S.: Redirecting network traffic toward a fake DNS server on a LAN. In: 2010 3rd IEEE International Conference on Computer Science and Information Technology (ICCSIT), vol. 2, pp. 429–433. IEEE (2010)
7. Djanali, S., Arunanto, F.X., Pratomo, B.A., Studiawan, H., Nugraha, S.G.: SQL injection detection and prevention system with raspberry Pi honeypot cluster for trapping attacker. In: International Symposium on Technology Management and Emerging Technologies (2014)
8. Drake, N.: Raspberry Pi Protection!. Future Publishing Ltd., Bath (2017)
9. Balani, S., Swathi, B., Shrestha, N.B.: Survey on home security surveillance system based on wi-fi connectivity using Raspberry Pi and IOT module. Udaipur Int. J. Adv. Res. Comput. Sci. (2018)
10. Feng, X., Babatunde, O., Liu, E.: Cyber security investigation for Raspberry Pi devices. Int. Ref. J. Eng. Sci. (2017)
11. McKemmish, R.: What is forensic computing? Trends Issues Crime Crim. Justice **118**, 1–6 (1999)
12. Hegstrom, K., McCoy, M., Creecy, J., Lord, W.: Use of forensic corpora in validation of data carving on solid-state drives. ProQuest Dissertations Publishing. Retrieved from http://search.proquest.com/docview/1865305312/ (2016)
13. Kiruthu, G., Rogers, M., Marshall, B., Rawles, P., Springer, J.: Digital forensic investigation of a Dropbox cloud-hosted shared folder. ProQuest Dissertations Publishing. Retrieved from http://search.proquest.com/docview/1836825640/ (2012)

Forensic Analysis on Kindle and Android

Archit Saxena, John Walker and Vedant Kulkarni

Abstract In this chapter, we conduct a forensic analysis of Amazon's Kindle Fire HD and report on our findings.

Keywords Walled garden · SPDY

1 Introduction

The worldwide tablet showcase developed to 6.5 million units in the final quarter of 2017 (4Q17), an expansion of 10.3% from the past Christmas season, as indicated by starter information from the International Data Corporation (IDC) Worldwide Quarterly Tablet Tracker. Development for the whole year stayed positive in spite of the fact that it hinted at abating as separable tablets developed 1.6% year over year in 2017, down from the 24% development in 2016. Nonetheless, a portion of the gradualness was ascribed to the dispatch rhythm of prominent gadgets like the Surface, which was off calendar, leaving more seasoned models on racks as shoppers and organizations laid in sit tight for item invigorates.

To date, a great part of the direction of the tablet market has been credited to Microsoft and Apple pushing their products in the U.S., "said Jitesh Ubrani, senior research investigator with IDC's Worldwide Quarterly Mobile Device Trackers." Notwithstanding, proceeded with accomplishment of this class depends on the eager-

A. Saxena
Department of Computer Science,
University of Texas at San Antonio, San Antonio, TX, USA
e-mail: archit.saxena@my.utsa.edu

J. Walker · V. Kulkarni (✉)
Department of Information System and Cyber Security,
University of Texas at San Antonio, San Antonio, TX, USA
e-mail: kulkarni.vedant21@gmail.com

J. Walker
e-mail: doc.walkerjhjr@gmail.com

© Springer Nature Switzerland AG 2020
X. Zhang and K.-K. R. Choo (eds.), *Digital Forensic Education*,
Studies in Big Data 61, https://doi.org/10.1007/978-3-030-23547-5_9

ness of other PC merchants to take an interest and all the more critically, buyers from different nations to receive the new shape factor over convertible PCs.

"After a concerning downturn in the last quarter of 2016 and first 50% of 2017, we are elated to see the separable market keep up another quarter of development," said Lauren Guenveur, senior research investigator for IDC's Devices and Displays group. "With the principal wave of Windows on ARM items anticipated that would start dispatching in the second quarter of 2018, we trust the detachable tablet class can possibly proceed with its development direction. A large number of these items are being presented at the top notch end of the market. What remains incredibly inadequate, and required, are solid players in the mid-fragment of the market."

Outside of the separable tablets, out-dated slate tablets kept on delivery in vast volumes achieving 43.1 million units amid the final quarter, and 141.7 million amid the year. These gadgets don't offer much as far as efficiency and have been to a great extent consigned to straightforward media utilization gadgets. With tight edges a few merchants are utilizing this frame factor as a gateway to their bigger biological systems or to advance their image inside more extensive contraption advertise. In the interim, end-client request has impeded fundamentally over the most recent couple of years, with 2017 declining 7.6% from the earlier year.

Apple's ipad kept up a strong lead in the occasion quarter driving development both through its lower-estimated 9.7-inch iPad and in addition its recently revived iPad Pro items. While the lower-value iPad has kept on driving a solid buyer update cycle, the moving center is to iPad Pro and its potential in the business and training product. With the main anticipated that tablet development would originate from these products, Apple's ongoing dispatch of "Apple at Work" demonstrates their promise to keep up its pioneer position.

Amazon.com could take the second position from opponent Samsung as the online mammoth offered soak limits amid the Christmas season. The minimal effort tablets have been very compelling as shopping indexes for Amazon's online customer facing facade and in late quarters the organization has made things a stride further by including its voice partner, Alexa, in the most recent tablets and by extending to new markets far and wide.

Samsung dropped to the third position behind Amazon. Shipments of its separable portfolio keep on rising, anyway these additions are exceeded by the decreases seen among its slate models. As its lower-cost Tab A and E flagship model is tested by merchants promising better esteem, Samsung will confront a specific test in redesigning its base to both a higher-evaluated and separable gadget.

Huawei's indifferent way to deal with the developing separable section does not offer a promising viewpoint for the organization. Be that as it may, the organization's solid image, forceful channel procedure, and incorporation of cell availability in its slate tablets has helped bond its rank in the main 5.

Lenovo's capacity to use its quality from the conventional PC business is beginning to satisfy as the organization's separable tablet business has developed in the previous year. In any case, the heft of Lenovo's tablets are still contained slates that offer extraordinary incentive at low costs.

The rest of the paper is organized as follows. Section 2 presents a literature survey of existing work on blood acquisition and distribution. Section 3 describes the system of systems design in blood supply management. Section 4 presents a model for balancing the blood supply in anticipation of an emergency event. Section 5 concludes this paper.

A. Kindle

The Fire HD, otherwise called Kindle Fire HD, is an individual from the Amazon Kindle Fire group of tablet PCs. The eight age family comprises of: 7″ (2012 model), 8.9″ (2012 model), 7″ (2013 model), 6″ and 7″ (2014 models), 8″ and 10.1″ (2015 models), 8″ (2016 model), 8″ and 10.1″ (2017 models), and 8″ (2018 model).

Hardware:

The Fire tablets highlight multi-contact touchscreen LCD screens. The original 7″ display contains a Texas Instruments OMAP 4460 processor, while the 8.9″ demonstrate utilizes an OMAP 4470 processor. All three models highlight Dolby sound and stereo speakers. The 7″ model's speakers are double driver, while the 8.9″ model's are single-driver. The gadget has two Wi-Fi receiving wires on the 2.4 and 5 GHz groups which use MIMO to enhance reception. The Fire HD likewise added Bluetooth network enabling clients to interface a variety of remote adornments including keyboards. The original models have a HDMI port, however this is absent from who and what is to come.

In June 2016, Amazon discharged a variant of the Fire HD 10 that has an aluminum outside rather than plastic like the other Fire tablets, and is accessible at indistinguishable cost from the plastic rendition.

Software:

The 2012 models utilize programming that presented client profiles for sharing among relatives and the capacity to put supreme breaking points on aggregate use or utilization of individual highlights, called FreeTime, and tracks the client's perusing pace to foresee when the client will complete a section or book. The OS depends on a variant of Android 4.0.3 (Ice Cream Sandwich). This does not permit utilization of Google Play, constraining the quantity of applications available for the Fire HD. Fire HD programming updates can be gotten OTA or from the help websites.

The Fire HD 7″ second era utilized Fire OS 3. Note that despite the fact that this rendition is known as the Fire HD 7″, it isn't the successor to the first Fire HD. This model is the successor to the Fire second era. The Fire HD models second era were refreshed to Fire OS 4.1.1, in view of Android 4.4.4, in Q3 2014.

The Fire HD 6″ and 7″ third era utilizes Fire OS 4 "Sangria", which highlights profiles so every client on the tablet can have their own settings and apps.

The Fire HD 8 and 10 fifth era utilizes Fire OS 5 "Bellini" and was discharged in late 2015. In September 2016, Amazon discharged menial helper Alexa for the 6th era Fire tablets.

The 2018 model of the Fire HD 8 has Fire OS 6 pre installed, which depends on Android 7. It additionally incorporates Alexa Hands-Free and the new "Show Mode", in which the tablet demonstrations like an Amazon Echo Show.

2   Rise in Popularity

Amazon's overall tablet shipments spiked 48% last quarter while the tech monster passed Samsung for second place in merchant rankings. Research firm IDC discharged new tablet showcase information on Tuesday. Amazon sold 7.7 million tablets in Q4 2017, bouncing in front of Samsung (7 million) surprisingly.

Amazon, which offered huge limits on its Fire tablets amid the Christmas season, had a 15.6% piece of the overall industry in the final quarter, trailing just Apple at 26.6%. "The minimal effort tablets have been very powerful as shopping indexes for Amazon's online customer facing facade and in ongoing quarters the organization has made things a stride further by including its voice collaborator, Alexa, in the most recent tablets and by growing to new markets the world over," IDC said.

Amazon's principle equipment items are the Fire tablets and Echo voice aide speakers; both come outfitted with Alexa. Tablet shipments generally dropped last quarter by 7.9% year-over-year. Here's a gander at shipments and piece of the pie for 2017 in general—take note of Amazon's 38% spike:

3   Different API

The accessibility of Amazon Appstore for Android SDK and In-App Purchasing API to our engineer network. The In-App Purchasing API empowers to offer advanced substance and memberships-, for example, in-amusement money, extension packs, updates, and magazine issues—for buy inside applications.

The In-App Purchasing API to make it simple to build client commitment and adapt the applications on Kindle Fire and other Android gadgets. With the Amazon Appstore for Android In-App Purchasing arrangement, can achieve clients with existing records who have just purchased applications, including a great many Kindle Fire clients. The basic, secure, and confided in 1-Click buy encounter is simple for clients to utilize, expanding transformation rates for buys inside your application. In addition, we planned our in-application buying (IAP) answer for be straightforward and simple to coordinate so you can be ready for action rapidly.

It's anything but difficult to begin from the Amazon Appstore Developer Portal, where you'll discover documentation, test code, instructional exercises, much of the time made inquiries, and that's just the beginning. Additionally be including methodologies and tips from in-application buying beta program accomplices, for example, Glu Mobile, G5 Entertainment and Storm8.

What do beta program accomplices say in regards to IAP on the Amazon Appstore for Android?

"Storm8 facilitated intimately with Amazon's group amid the underlying dispatch of its IAP beta test, and inside two months of incorporation, saw income develop by ten times from amazon's amusement," said Perry Tam, CEO and fellow benefactor at Storm8, maker of recreations, for example, Restaurant Story, Bakery Story, Farm

Story, and Fashion Story. "Instantly brought over extra Storm8 diversions, and in about fourteen days, not exclusively did the income keep on developing, however had four of the best five free applications in the Amazon Appstore. With the enormous introductory achievement, certainly plan on proceeding to put resources into the stage and can hardly wait to bring extra Storm8/TeamLava diversions to Kindle Fire and Amazon clients."

"Amazon's in-application acquiring arrangement made an incredible path to diminish rubbing and drive more income from the diversions, as a large number of individuals as of now, Amazon accounts," said Michael Ritter, senior VP Licensing and Distribution at Social Gaming Network, producer of Warp Rush, Dress Up! Mold, Bird's the Word, and Night of the Living Dead Defense. "Arouse Fire as of now has a very much incorporated retail facade and commercial center to convey versatile diversions. By empowering in-application buys can be more adaptable in evaluating. Amazon discharge free diversions, give updates, and improvements, and keep on adapting."

"Discovered that by offering an item with IAP, as opposed to other adaptation types, transformation rates went up as did the amazon's income, for every title premise," said Larissa McCleary, chief of promoting at G5 Entertainment, Inc., maker of Virtual City Playground and Mahjong Artifacts. "Despite the fact that experience on Amazon has dependably been incredible, we are excited now that IAP is accessible. In the end, assuming an ever increasing number of engineers take an interest, figure out what encounter significantly higher change rates, since players will be more comfortable with what IAP is and how it capacities, making the play more captivating and interesting."

4 Security Through Obscurity

With the pattern toward a very portable workforce, the procurement of handheld gadgets tablets, for example, Personal Digital Assistants (PDAs) and PC tablets is developing at a consistently expanding rate. These gadgets offer efficiency instruments in a reduced frame and are rapidly turning into a need in the present business condition. Numerous makers make handheld gadgets utilizing a wide scope of equipment and programming. Handheld gadgets are portrayed by little physical size, constrained capacity and preparing power, limited stylus-arranged UI, and the methods for synchronizing information with a more competent notepad or work station. Normally, they are outfitted with the capacity to impart remotely over constrained separations to different gadgets utilizing infrared or radio signs. Numerous handheld gadgets can likewise send and get electronic mail and access the Internet.

While such gadgets have their constraints, they are regardless amazingly valuable in overseeing arrangements and contact data, inspecting archives, relating through electronic mail, conveying introductions, and getting to corporate information. Additionally, due to their generally ease, they are getting to be pervasive inside office situations, frequently bought by the representatives themselves as a productivity help.

Lamentably, a few noteworthy issues, over the utilization of such gadgets, including the accompanying things:

1. As a result of their little size, handheld gadgets might be lost, left unattended, or stolen.
2. Client confirmation might be crippled, a typical default mode, uncovering the substance of the gadget to any individual who has it.
3. Regardless of whether client validation is empowered, the verification component might be frail or effectively evaded.
4. Remote transmissions might be caught and, if decoded or encoded under a defective convention, their substance made known.
5. The simplicity with which handheld gadgets, the amazon kindle can be interconnected remotely, joined with powerless or no validation of the gatherings included, gives new roads to the presentation of infections or different sorts of noxious code, and furthermore different types of assault, for example, a man-in-the-middle attack.
6. The mobile agent information flow raises a few security concerns, which are one of the primary snags to the far reaching use and adaptability of the new innovative approach. Security issues include: Authentication, Authorization, Confidentiality, trusting the third party vendor, non-reputation, asset monitoring and denial of service attacks.

5 Walled Garden

According to the wikipedia walled Garden is defined as "A closed platform, walled garden, or close ecosystem is a software system where the carrier or service provider has the control over applications, content and media, and restricts convenient access to non-approved application or content."

US online advertisement incomes at Amazon will dramatically increase this year, eMarketer gauges, moving it up the positions past Oath and Microsoft to take the No. 3 position behind Facebook and Google.

US promoters will burn through \$4.61 billion on Amazon's stage this year, representing 4.1% of all computerized advertisement spending in the nation. Note that a lot of that development is because of a bookkeeping change Amazon made that influenced evaluations for 2018 and later, prompting the development spike this year.

All things being equal, the conjecture builds eMarketer's gauge of advertisement incomes at Amazon by about 10–12% through the span of the year, on account of more grounded than-anticipated natural development.

Also, that expansion is plainly still vigorous. After 2018's counterfeit development spike, eMarketer expects increments of over half every year through no less than 2020, when Amazon will appreciate 7.0% of all US computerized advertisement spending.

For over a year currently there's been an enduring mumble about Amazon's infringement on the region of the Facebook-Google duopoly. This year, 57.7% of

US computerized promotion spending will go to one of those two stages. Amazon might be in third place, yet 4.1% is a long ways behind Facebook's 20.6% or Google's 37.1% of piece of the pie. In 2020, Amazon's 7.0% offer will contrast and Facebook's 20.8% and Google's 35.1% of US advanced advertisement spending. So while the hole might limit, the duopoly column still stands.

Amazon has the elements for a genuinely powerhouse advanced advertisement business, the way to which is accomplishing something other than offering promotion impressions. It's difficult to state the amount of Amazon's business originates from offering promotions all alone properties versus encouraging advertisement exchanges over the web. Be that as it may, with the promotion tech assess representing a huge offer of all showcase spending, there's no chance to get for a over the internet advertisement merchant—not in any case Google or Facebook—to split the most elevated income rankings without profiting from those income streams too.

What's more, Amazon is currently making it less demanding to purchase advertisements on its stage. Numerous purchasers have grumbled that Amazon's promotion items were befuddling and hard to purchase. On September 5, Amazon declared that all promotion purchasing, and detailing would fall under another "Amazon Advertising" standard and be completely combined before the year's over, with new names for some contributions.

Real changes include:

Amazon Marketing Services (AMS, Amazon's suite of expense per-click (CPC)-based promotion designs including Sponsored Products), Amazon Media Group (AMG, the specialty unit that sold showcase publicizing on Amazon properties and Kindles) and Amazon Advertising Platform (AAP, now known as Amazon DSP) are getting to be Amazon Advertising.

Feature Search promotions—an expense for every snap (CPC) arrange that includes different items on query items pages—is being renamed Sponsored Brands. Supported Products advertisements and Product Display Ads, other CPC groups, will keep similar names.

eMarketer's most recent report, "Amazon Advertising 2018: The New No. 3 US Digital Ad Firm," investigates how advertisers and media purchasers are intending to utilize Amazon for promoting, and how Amazon's gathering of people information and item look abilities offer a point of separation from Facebook and Google.

6   How We Got Here

The initial topic of our research was to analyze data from an Amazon Fire HD Tablet. We wanted to know what, if anything was different in an Amazon Fire device that differs itself from a non Amazon device. We acknowledge that other third party partners are able to modify Android to their liking [1] and do so, but Amazon Fire is not marketed as Android but its own operating system. In addition Fire OS is a start into the "walled garden" approach to ecosystem the most famous of which is

Apple's iOS. iOS however, is an open and shut case with forensics tools matured and well developed. These reasons were the primary factors in wanting to learn if modern Fire devices were fundamentally different from the android source that they are forked off of.

A refurbished Amazon Fire HD 8 (7th Generation) was bought for the analysis of this research. This device was chosen for its, cost and for its age, as it was released over a year ago in July 2017, and the newest edition of the Fire HD 8 would not be released until October of this year [2]. Lastly it is interesting to note that the new HD 8 model seems identical in hardware and the only major change listed is the inclusion of the newest Fire OS [2]. On initial inspection this version of the tablet came with "special offers" a function of Fire tablets that display ads on screen. Another question was added to our investigation list: what kinds of ads were stored on the device, as ads would display without a wifi connection, and do they relate to the user's activity? Being a wifi enabled device we found that the Fire HD 8 came with a maps application. Further research on this revealed that Amazon have licensed Nokia's map API [3], and the map application included something called "itineraries", and wanted to pull from contacts to add people to an itinerary. Lastly our literature review noted that the Silk web browser would be of forensic interest because it utilized Amazon's Elastic Compute Cloud (EC2) services for faster browsing [4]. Our final topics would include difference in Fire OS to Android, analysis of the maps application, and analysis of the Silk Browser.

It was quickly learned that we would need to get root permissions on the device in order to extract data. Mobile Phone Examiner (MPE) was not equipped to deal with this model and was not able to extract the data we needed. While we could gain simple shell access with the Android Debugger (adb), we would not be able to access the relevant system partitions for the collection of data. Currently there only exists a hardware modification to gain root access. This involves gaining access to the motherboard of the device, attaching wires to certain pins to an SD card adapter, and using an SD card reader, manipulating the Amazon file system mounted from another OS [5]. We did attempt this but our tools were limited and ultimately after many destroyed SD card adapters, and snipping the wires connecting the motherboard to the speakers we gave up attempting to hard mod the Fire HD 8. Currently the search for a software root is ongoing, and there currently exists a way to load an additional OS side by side Fire OS but not an actual root for the device itself [6, page 94]. Knowing that no root exists over a year later and with new models coming out, a software root progress for the Fire HD 8 may not materialize.

We decided to pivot into basic android forensics. We still wanted to do a research paper with hard data. We decided that gaming apps should be our focus. Forensically, gaming apps could prove useful for an investigator for a number of reasons. First it could show.

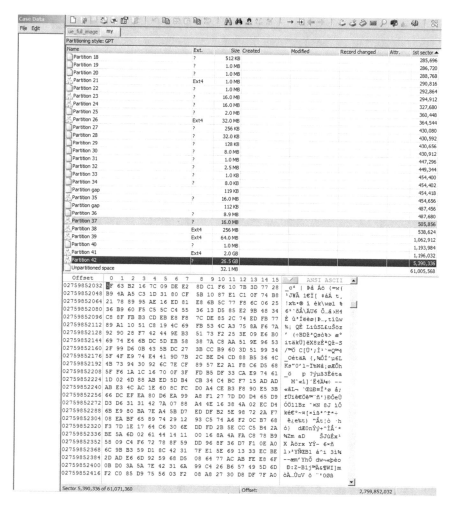

when someone was playing, many gaming consoles have their own messaging services which could allow for additional communications and could help build a profile. It could show alias names or purchase history, and verify console by ID to the player. Some applications can interact with the games directly instead of being an ancillary application like in game encyclopedia. In order to analyze this data we looked at the Xbox Live application for Android on a Google Nexus 6 manufactured by Motorola.

MPE was not as helpful as we hoped. It did not list the Nexus 6 as a device it would be able to read. We did attempt other configuration that might have been similar to the Nexus 6 but we did not know if we could trust that MPE had extracted a complete image or not. Therefore we needed root access again and this time was more fruitful. There did exist a software root but in doing so we would erase all the phone's data [7]. It was determined it was acceptable to acquire root this way because we should be able to use the same app with the same account and relevant data, if any should be downloaded upon the apps initialization.

Rooting the device was successful but we ran into two issues regarding extracting the relevant data. First was an issue with time. Our first way in extracting an equivalent binard dd of the hard drive took enormous amounts of time. This involved piping the dd command over a socket connection using adb, through the USB interface. A 32 GB image took over two hours to download. If the USB connection was interrupted the download would stop. We also tried over a wifi connection but this too was not quicker. Our second issue came when we attempted to view user data. The userdata partition on the Nexus 6 is encrypted by default through the kernel, but thankfully users worried about the performance of encryption and decryption have found ways around it [8]. This, like rooting the device again required to flash custom boot images and kernels. This time however, we were able to flash a custom recovery image which proved useful later and allowed us to flash the images for the kernel without encryption. During this time we did temporarily flash the wrong images and could not boot an operating system as the phone would only boot into fastboot mode; we note this as one of the many challenges that anyone might face in forensically recovering data.

With an unencrypted partition and a recovery image with more functionality than the phone's stock model, we scripted out the extraction of every partition in the Nexus 6 file system and named them accordingly to their names when mounted as found in /dev/block/platform/xx_xxx.0/by-name folder. Of note there were two large partitions, system and userdata, at roughly 2 and 27.8 gigabytes each. These were the ones we focused on in our brief analysis when looking for relevant forensics artifacts toward our topic.

We found the apk for the Xbox live companion application as well as where it stores its settings. However after analyzing the apk and not finding anything relevant like a database in the application's data storage location, we decided that we may not have a winner on this topic and analysis may not be done easily with file system analysis. The apk revealed javascript and compiled xml files among its assets. Very little information was actually accessible in the data files themselves save for some identifiers such as the Xbox gamertag and cached image files of games in the corresponding library. Our initial assumption is that the application is merely a front end for javascript and web queries to Microsoft's Xbox Live back end but can not confirm with some more reverse engineering of the apk compiled java byte code.

Finally with time running out in the assignment and no relevant data from third party gaming apps, the decision was made to conduct a meta analysis on anything and everything relating to the Amazon Fire ecosystem. Given our initial literature review not given much in the way of Kindle forensics information, we extend the research into as much published information into Kindle Fire OS, including but not limited to the Fire TV, Echo devices, and Alexa artifacts.

- Email

    – Autopsy—None
    – MPE+—Yes (62)

- Message, Phone Contacts, Images

  - Autopsy—121 Contacts, 1 text Message.
  - MPE+—121 Contacts

- Deleted Files

  - Autopsy—Yes, plethora of files after many phone wipes and restores
  - MPE+—Yes, plethora of files after many phone wipes and restores

- User Data

  - Autopsy—Yes, but very little remained after multiple phone wipes.
  - MPE+—Yes, but very little remained after multiple phone wipes.

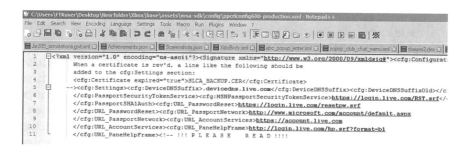

7   Literature Review

Mobile phone operating systems have converged to two main players. Apple's iOS and the open source Android maintained by Google. Apple and Google fourth quarter 2017 report by IDC (International data corporation) make up 99.9% of the mobile market OS [9]. Android has an interesting UI where anyone can use its source code for their own mobile devices. This is precisely what Amazon has done with their FireOS and furthermore Amazon has made their OS proprietary. Amazon's market share in tablets has grown to 10.2% at the end of 2017 which is still dominated by Apple devices [9]. Proprietary software and operating systems deviating from the standards because of the cheap and satisfactory result by kindle fire hd tablets, therefore Android, may be the last frontier of traditional mobile forensics for some time.

Amazon developed the kindle fire which is known as fire tablets built with the collaboration of Quanta Computer (Taiwan-based manufacturer of notebook and electronic hardware). The Kindle first edition was released in November 2011, and has had many iteration and generation to follow it to the present day [6]. The thesis on "An exploratory forensic acquisition and analysis of digital evidence on the amazon kindle by Marcus Thompson was focused on the fifth generation of Kindle E-readers and focused on the question if forensic analysis was viable with the devices. What is notable is that by then the Kindle Fire had been released two years prior, and the Kindle E-readers he chose to analyze where two to three years old as well. Thompson was able to conduct basic file carving techniques by populating the devices with files with known hashes, deleting them and retrieving them through FTK and comparing the hashes again. No other analysis was performed however he notes these limitations and shortcomings which is an indicator of future study [10]. Thompson also had a blog referenced regarding a simple overview of Kindle Forensics in his blog during 2011. His blog is a bit more technical and revealing but is also much more informal and doesn't fully adhere to a sound forensics investigation practices but rather doing this for general purpose research. This entry reveals Kindle internals, file system layout and storage. [11].

Building Thompson's blog, Iqbal et al. [1] compiled a literature review for mobile forensics specifically targeted for their research into the modern, more advanced Kindle Fire. Their compiled knowledge covers a foray of early 2000's mobile forensics to Amazon specific forensics which were mostly blogs at the time and ending with harder research of Android forensics including imaging techniques, memory dumping and file system analysis, specifically YAFFS2. Their work concluded with a more detailed forensic overview of the Amazon Fire tablet and highlighted specific areas of interest to a forensics investigator, where they are found and what to expect to be found in those locations. One of the authors later went on to publish another research paper with another team but similar purpose: to outline the acquisition and analysis of a Kindle Image for forensics investigation. They did successfully identify new ways to acquire a forensic image and left about the same amount of detail as Iqbal et al. [12].

While the previous articles lay foundational work for Amazon Kindle Fire forensics there are still knowledge gaps within the walled garden ecosystem that Amazon is building. One of which mentioned in Iqbal, Baber et al. was the Silk Browser. The Silk browser offers new challenges for the forensics investigator as browsing is now "split" between the device and the Amazon EC2 cloud as well as being proprietary and non open source [13]. A recent study [14] of Amazon web services posits the question of acquiring and analyzing Amazon EC2 data for future research. The authors note that EC2 imaging was done in a previous study [15] but that still did not eliminate the challenges of acquiring forensically sound images of an EC2 instance, and required trust as well as online persistence. Because of the reliance on EC2 the Silk browser may have differing artifacts available on disk or in memory, especially since Silk applies machine learning to pre-fetch pages a user is likely to visit next. Otherwise it is noted that Silk can run in non-cloud mode, which implies it renders everything locally and normal android browser forensics apply, though it is unknown if this is an option in later models [13]. Of note there appears to be a tool that was created for research purposes for additional browser forensics. Christopher Neasbitt et al. created a record and replay forensics application entitled WebCapsule. It heavily integrates with Blink, the Chromium web engine which grants it a degree of portability which the authors do try in mobile browsers, though not specifically Silk [15].

Briefly Amazon's Fire Tablets come with Special Offers, which are expensive lock screen ads [16]. It is unknown where they are stored, how often they get refreshed or pushed and how targeted they are. However what is known is where the setting is to turn them off [17]. There may be forensic evidence within the special offers ads or Amazon's recommendations of previously bought goods and entertainment.

Amazon has its own maps API [18] which is licensed from Nokia [3]. This makes it a different beast than Google Maps or Apple Maps. By now previous research is in the basic vernacular and Google Maps forensics is known and taught in books discussing android forensics, though some are more thorough than others [19, 20]. Given map apps on the Amazon App store have none of the other big players (Apple, Google, Microsoft, Yahoo even) and excluding the failed Fire phone, a Maps App for a tablet is likely to have less features than a fully-fledged navigation app such as Google Maps.

Iqbal, Babar et al. was concerned with third party forensics on the Amazon Kindle. The lion's share of Apps on the Amazon App store is games with over 200,000 out of at least 600,000, (the that number is claimed to be 600,000 apps on the app store we know there is at least 600,000) [21]. Comparison of third party apps likely boils down to what is popular and is on Amazon's App store, Google's Play Store, or Apple's App Store. Specifically it is known which Amazon services correlate to Google Play Services allowing for easier development of cross functional apps [22]. It is unlikely that third party apps will differ on the Amazon Fire and forensic analysis will be needed on a relevant app by app basis.

The paper on "Analysis of Privacy of Private Browsing Mode through Memory Forensics" [23] by Ahmad et al. intends to use a framework to capture and analyze the memory that is associated with private browsing. The paper uses evidence that how using that particular framework will help the investigators and the incident responders reconstruct the past browsing history in case of an incident, so in our research we intend to conduct forensics on silk browser, the methodology used in the paper does not conduct forensics on the amazon's silk browser, but instead they have used google chrome, IE, Safari and Firefox.

According to a magnet forensics [24], published the results in February, 2014, forensics was performed on kindle fire and kindle fire HD through IEF (Internet evidence finder). They found amazon fire tablet run on a customizable android version of Ice Cream Sandwich. IEF can acknowledge the kindle image and supply the examiner with all the potential artifacts related to it. Several of those artifacts hold on within the same databases as vanilla robot installation, however you may notice couple of variation. The kindle fire has its own browser silk which uses the Amazon AWS and SPDY protocol to boost the content instead of the traditional HTTP. Other artifacts in the kindle was found in the same directory and location as they would be in android version 4.0+. The kindle fire forensic analysis with IEF found the similarity with the android version 4.0+ and was able to collect all the artifacts successfully.

Paper on "Kindle Forensics: Acquisition and Analysis" by Peter Hannay illustrates that digital book gadgets, for example, the Kindle are gathering expanded enthusiasm from the scientific network as they turn out to be progressively mainstream. The included cell capacity of the Kindle particularly may make it a contender for evil purposes, as the there is no information cost related with the worldwide information benefit [25]. Notwithstanding information usefulness the incorporation of an application structure and advancement pack in beta release discharge will just prompt expanded utilization of the item for purposes that were once met by the conventional figuring worldview. The underlying endeavors of the measurable network have concentrated on obtaining of just a segment of the inner storage of the gadget as this territory is promptly open as a USB mass storage gadget. This paper has gone past the current strategies and gave an instrument to the procurement of the entire inward NAND memory and examination of same. All together for this outcome to be accomplished anyway a few information must be composed to the gadget and in doing as such there is the likelihood of information being overwritten. Anyway beside intrusive equipment based securing there are no present known procedures that would take into account finish procurement without this methodology. Investigation into little and inserted gadget criminology is continuous, with expanded spotlight on total securing of every applicable datum from these frameworks, including storage, memory and information put away on individual microcontrollers.

8 Findings

When it comes to Amazon Kindle forensics what surprised us most was the lack of readily available information and research. Using the UTSA's library to search for relevant information we discovered there were not a significant amount in their searches. In addition we did have to filter out some other information from other sources like magazines and litteral fires in the amazon rainforest of Brazil. Barring these it was a bit of relief that we were not the first to have a desire to analyze Amazon's Fire tablets

The shortlist of compiled findings shows the bare bones of information that a forensic investigator would want for Amazon Kindle devices for the times. Specifically it was interesting to note that research started with the e-readers themselves and not the more advanced Fire HD tablets. Marcus Thompson's research went into the basics of file carving and file extraction on the original e-readers, when Fire OS was not even conceived and the devices ran a basic Android operating system, but written and researched at a time just as they were evolving. Reasons for choosing as many e-readers as he did for his research might have been cost as the newer tablets were released as well as accessibility as rooting devices would be needed. Even though he only touched file carving and file acquisition it served as a basis that Kindle devices were not security minded and originally not challenging to the forensics investigator.

Iqbal, Iqbal, Dr. Guimaraes, Khan, and Al Obaidli, had earlier analyzed a Kindle Fire in 2012. Their research along with Iqbal, Al-Obaidli, Baggili, and Marrington, continued into Kindle Fire as they become more solidified tablets and an Amazon brand with the Kindle Fire and Kindle Fire HD respectively. They are lumped together because their papers conclude similar findings, not least because they share two of the same authors. What was discovered and ultimately concluded was a basic path or blueprint for future forensic investigative work. However the versions that both of these researchers analyzed was based off of Android and not labeled Fire OS proper. I mention this because their findings in "Amazon Kindle Fire from a Digital Forensics Perspective" attempt to find relevance in finding a contacts database within the operating system and user data of an e-reader. With no specific focus on tablets and not being rolled into a unified operating system, it is worth questioning how much vestigial folders, functions and applications were shipped with the first generation Kindle Fire tablets, that would normally be supported in a mobile phone. Regardless of the applicability of the files they found, they do provide a solid base for starting forensic investigations on an Amazon Kindle Fire devices. As a side note, Jacob Horowitz also provided some insight into the third generation of Amazon Fire devices, but published his methods in a slipshod informal manner as a blog post. There were hardly as many findings as the research papers but still was delicate with regard to having a forensically sound image to investigate. Largely his work did not build upon the research papers but rather verified similar findings exist in the Kindle Fire HDX which was running Fire OS 3.

All three of the previous researchers needed some kind of root access. Given our own troubles with acquiring root access with our own Kindle device it should be a given that root access is almost necessary to gain forensically sound evidence and without it would be a difficult and uphill battle to climb. Though we were not able to the researchers had a number of ways in which they were able to gain root access. Primarily they all had access to a software root that was mostly embedded and a fault of the operating system. However with the Kindle Fire HD it is interesting that they used a modified USB cable to gain communications with the underlying hardware to push a custom boot image to preserve user data. The use of hardware devices like this seems an odd design choice but a bountiful one to the forensics investigator. Such cables are more easily to modify than the hardware mod presented before us with our attempts at the Kindle HD 8 7th generation. Their other exploit was also novel in that it modified settings to make the device boot under the assumption it was emulated under QEMU when it was not. This allowed for the extraction of userdata partitions without worry or potentially overwriting wrong images.

Our research has large gaps in years between 2014 and 2017. Outside of additional blog or forum posts that may or may not be fully relevant to forensics, there is a large gap of research. At or around this time Amazon started shipping Kindle Fire HD devices with a fully branded and changed Fire OS. It could be the case that finding and waiting on root exploits to analyze the internals of the latest and greatest models was not worth it, or their market share was still not great enough to warrant additional research based on the previous findings and similarities that an Amazon Fire Tablet isn't that much different from any other Android tablet. There is hope however in that in 2014 Magnet Forensics a provider of investigative technology products, included a brief video that one of their products, Internet Evidence Finder, would analyze Kindle Fire images. However, in their video they had already created a base image to show in their tutorial but only goes on to talk about internet and browser artifacts parsed through the Silk Browser databases. Of note this article verifies that the Silk browser's location has not changed and still resides in/data/data/com.amazon.cloud9. This is also an easy and accessible tutorial that tells us that Amazon Kindle forensics aren't being ignored or are too obscure and are being included in professional products. Today, the Mobile Phone Examiner + program has the ability to analyze some Amazon Kindle devices.

The final two papers on "Digital forensic approaches for amazon alexa ecosystem" and "An Investigation of cloud forensics and the Artifacts of Amazon Web Services" analyze something more relevant in our moving times. It is no surprise that Amazon created Internet of Things (IoT) devices with the Amazon Echo, Echo Dot, Echo Spot, and Echo Show. These, along with their modern Fire HD tablets come equipped with Alexa, the smart voice assistant. Hyunji Chung, Jungheum Park, Sangjin Lee published an amazing research paper on Amazon's Internet of Things devices which was a treasure trove of potential forensics applications. Their research was aimed at these connected devices wondering what if anything are stored. Because of this ecosystem had become so overbearing so quickly, the researchers created their own toolkit to analyze IoT artifacts, and then created an extensible framework for IoT devices. Using their tool they made their own modules based on what they found. They looked at the connected ecosystem that Amazon has created and categorized four areas of interest for finding artifacts, Cloud, Client, Network and Hardware. Hardware was left relatively untouched but the cloud client and network artifacts revealed much more. Their main discovery yielded the hidden API calls that Echo devices can query revealing information about device usage including name, emails, and additional linked accounts. Many of the queries came with UNIX timestamps which was noted could be forensically interesting and may be essential data a user is not able to manipulate even with access to the Echo devices.

The last paper on "An Investigation of cloud forensics and the Artifacts of Amazon Web Services" to look out was published a few months ago this year. In it Ian Gress looked at what forensic information could be acquired when using Amazon Web Services (AWS). He conducted browser forensics in a VM to see what kind of information the browser would leave behind when using the AWS Simple Storage Service (S3) as well as what information Amazon's AWS security tool CloudTrail which tracks a user's activity. His findings weren't specifically related to mobile services as he conducted his research in a windows VM in addition he didn't test many browsers for variation and stuck to chrome. As Silk browser uses Elastic Cloud Computing (EC2) to speed up browsing and Alexa devices can be queried through API's this paper seemed relevant to include in our analysis even if not directly tied to an Amazon Fire device.

Lastly there appears to be some lag in what research achieves toward the cutting edge. Every single paper, blog and article seemed to be behind the times with regards to the devices they were analyzings. While research takes time and allows for others to build upon their work there seemed to be a lack of haste when it comes to new devices. Even in our own case with the Amazon Kindle HD 8 7th generation there was no software root almost two years after its release. Given that we still lack a software root, and all the other papers seem to have some technical problem solved such as rooting a device or writing an extensible IoT framework for current and future research, it is clear that forensics investigation and the research that goes into it takes time.

9 Conclusion

Amazon's Fire tablets run Amazon's own "Fire OS". Fire OS uses google android code but they do not use any one of the google services such as google play etc. This is what that implies, and how precisely they're extraordinary. It's not by any means right to state that Amazon's Fire tablets run Android. In any case, in another sense, they do run a great deal of Android code. All the applications you'll keep running on a Fire tablet are Android applications, as well. You'll be utilizing Amazon's very own applications—the Silk Browser rather than Chrome, for instance. Kindle Silk browser uses a cloud based Webkit program. It utilizes the Amazon AWS to accelerate content for the program by running in parallel and exchanging the information through the SPDY, not HTTP protocol. SPDY protocol are used for: faster browsing the web, reduced latency, minimize deployment complexity, concurrent HTTP request, reduced bandwidth, better security etc.

Reserving and proxying on EC2: To accelerate perusing, Silk utilizes EC2 to complete a portion of the elements of an intermediary server and a store. The customer bit of Silk passes a URL to the EC2-facilitated divide, which at that point goes out and gets the majority of the parts fundamental for rendering the URL—HTML, CSS, pictures, Javascript, and so forth.. Since huge numbers of these assets are originating from destinations facilitated on Amazon's cloud, EC2 will have the capacity to stack them rapidly and, however Amazon didn't make reference to this in the declaration, without paying any sort of transport costs since that inside AWS movement doesn't go over people in general Internet and is basically "free" to Amazon.

The EC2 part of Silk would then be able to advance these assets for the specifics of the customer that asked for them; so if a picture is too substantial, for example, EC2 will resize it to fit the customer's screen goals. There's likewise a machine-learning segment to this storing/intermediary usefulness, where the EC2 segment will screen a customer's perusing examples and utilize that information to anticipate which pages the customer is probably going to stack straightaway. EC2 will then pre-get the anticipated page's parts and start take a shot at them, with the goal that the total page can be immediately conveyed to the client by means of single server-to-customer interface.

Rendering pages on EC2: When the segments of a page have been gotten and reserved on EC2, the page must be rendered for presentation in the customer's program window. Contingent upon the measure of load on the customer and the customer's system conditions, Silk can hand off a large portion of the real rendering pipeline to EC2.

The Silk video that Amazon posted on its bloglists the accompanying as parts that can be progressively given off to EC2 with the end goal to accelerate perusing: Networking, HTML, CSS, Collections, JavaScript, Marshaling, Native OM, Formatting, Block Building, Layout and Display.

For the normal individual, the huge contrast between a customary Android tablet and Amazon's. There are different contrasts, obviously. Amazon doesn't roll out it conceivable to improvement the launcher as you typically can on Android gadgets, so you'll be utilizing Amazon's home screen involvement. Amazon's home screen experience can demonstrate a lattice of applications, yet it additionally indicates you recordings, music, and digital books from Amazon. The home screen even contains Amazon's shopping site, making it simple to purchase more stuff—and give Amazon more cash.

Fire OS has a decent, kid-accommodating "Kindle FreeTime" include that can be joined with a membership for access to a great many child well disposed instructive applications, books, motion pictures, and TV appears. Amazon even offers a Fire Tablet structured particularly for children that groups in various application and includes a pleasant, "kid-friendly" feature. These "kid friendly" are parental-control are one of Fire OS's more extraordinary highlights.

Yet, what does the distinction extremely mean? All things considered, in the event that you simply need an economical tablet for perusing the web, experiencing messages, and watching recordings, there isn't that enormous a distinction. In the event that you need the whole biological system of Android applications without going through the motions, you should need to get a more common Android tablet.

That is Amazon's incentive, all things considered. You can get a modest, $50 Kindle Fire tablet—however you'll need to utilize Amazon's appstore and application rather than Google's. Amazon would like to make more cash off you in computerized deals. The least expensive variant of the tablet even ships with lock screen ads, and you need to pay some additional on the off chance that you need to expel them.

Yet, Android is additionally an open source venture. The open source venture is known, sufficiently just, as the Android Open Source Project (AOSP). The AOSP code is authorized under a lenient open-source permit, and any producer or engineer can take the code and utilize it for what they need.

Google Mobile Services isn't a piece of the Android open source task, and bunches of things that individuals consider as "Android"—including the Google Play Store and the majority of Google's services—are excluded in Android. They're authorized independently.

Digital book gadgets, for example, the Kindle and the fire hd tablets are gathering expanded enthusiasm from the forensic analysis as they turn out to be progressively prevalent. Kindle cellular capability particularly may make it a contender for accursed purposes, as the global service data costs are negligible. Notwithstanding information usefulness the consideration of an application structure and advancement unit in beta discharge will just prompt expanded utilization of the item for purposes that were once met by the customary computing paradigm.

Acknowledgements Dr. Raymond Choo, for guiding, teaching, and giving us the chance to experience and appreciate this research.

References

1. Iqbal, A., Alobaidli, H., Baggili, I., Marrington, A.: Amazon kindle fire HD forensics. In: Digital Forensics and Cyber Crime: Fifth International Conference, ICDF2C 2013, pp. 39–50, Moscow, Russia (2014). https://doi.org/10.1007/978-3-319-14289-0_4
2. https://www.amazon.com/b/?ie=UTF8&node=6669703011&ref_=fs_ods_fs_tab_cp
3. Wauters, R.: https://thenextweb.com/mobile/2012/09/17/nokia-confirms-amazon-licensed-location-platform-maps-geocoding/, 17 Sept 2012
4. Gress, I.: An investigation of cloud forensics and the artifacts of amazon web services (Order No. 10932214). ProQuest Dissertations & Theses Global. (2102542663). https://login.libweb.lib.utsa.edu/login?url=https://search-proquest-com.libweb.lib.utsa.edu/docview/2102542663?accountid=7122
5. https://forum.xda-developers.com/hd8-hd10/orig-development/root-hardmod-root-amazon-fire-hd-8-7th-t3851617
6. Amazon Inc. Amazon.com; Which Fire Tablet Do I Have?. https://www.amazon.com/gp/help/customer/display.html/?nodeId=201263780
7. https://forum.xda-developers.com/nexus-6/general/how-to-nexus-6-one-beginners-guide-t2948481
8. https://joyofandroid.com/how-to-disable-encryption-on-nexus-6/
9. Framingham: Idc.com; Detachable Tablets Return to Growth During the Holiday Season as Slate Tablet Decline Continues, According to IDC. https://www.idc.com/getdoc.jsp?containerId=prUS43549518 (2017)
10. Thompson, M.: An Exploratory Forensic Acquisition and Analysis of Digital Evidence on the Amazon Kindle. Purdue University (2014). https://search-proquest-com.libweb.lib.utsa.edu/docview/1616662019/?pq-origsite=primo
11. Thompson, M.: Introduction to Kindle Forensics. In: Practical Digital Forensics, 5 Sept 2011, [Cited: 27 Mar 2012]. http://practicaldigitalforensics.blogspot.com/2011/09/introductionto-kindle-forensics.html
12. Horowitz, J.: Kindle Fire HDX Forensics, 14 Apr. 2014, kindlefirehdxforensics.blogspot.com/
13. Stokes, J.: Amazon's silk is more than just a browser: It's a cloud OS for the client. Wired, Conde Nast, 1 Sept 2011. www.wired.com/insights/2011/09/amazon-silk/
14. Dykstra, J.: Digital forensics for IaaS cloud computing. (2012). https://digital-forensics.sans.org/summit-archives/2012/digital-forensics-for-iaas-cloud-computing.pdf
15. WebCapsule: Towards a Lightweight Forensic Engine for Web Browsers. 2015, Acm, pp. 133–145 (2015). https://www.longlu.org/downloads/ccs2015.pdf
16. Remove Amazon lockscreen ads w/ ADB: https://forum.xda-developers.com/hd8-hd10/development/remove-amazon-lockscreen-ads-adb-t3781218
17. Del Rey, J.: http://adage.com/article/digital/amazon-sell-ads-kindle-fire-screen/234830/, 17 May 2012
18. Amazons Developer Blog: https://developer.amazon.com/blogs/post/Tx14BH5AW0NG41K/Amazon-Maps-API.html, 16 Sept 2012
19. Tamma, R., Tindall, D.: Learning Android Forensics. Packt Publishing Ltd. (2015)
20. Hoog, A.: Android forensics investigation. In: Analysis and Mobile Security for Google Android. Elsevier (2011)
21. Amazon App Store Games: https://www.amazon.com/b?ie=UTF8&node=2478844011
22. Amazon Developers Reference: https://developer.amazon.com/docs/fire-tv/fire-os-overview.html
23. https://pdfs.semanticscholar.org/6a0b/57375bf2d6059c051756365af84d9890e0ee.pdf
24. https://www.magnetforensics.com/mobile-forensics/forensic-artifacts-on-a-kindle-fire/
25. Hannay, P.: Kindle forensics: acquisition and analysis (2011)
26. Egham, U.K.: Gatner.com; Gartner Says Worldwide Sales of Smartphones Recorded First Ever Decline During the Fourth Quarter. https://www.gartner.com/en/newsroom/press-releases/2018-02-22-gartner-says-worldwide-sales-of-smartphones-recorded-first-ever-decline-during-the-fourth-quarter-of-2017, 15 Feb 2017

27. Iqbal, B., Iqbal, A., Guimaraes, M., Khan, K., Al Obaidli, H.: Amazon kindle fire from a digital forensics perspective. In: Iqbal, B. et al. 2012 International Conference on Cyber-Enabled Distributed Computing and Knowledge Discovery (CyberC), pp. 323–329. IEEE (2012)
28. Chung, H., Park, J., Lee, S.: Digital forensic approaches for Amazon Alexa ecosystem. Digit. Investig. **22**(Supplement), S15–S25 (2017). ISSN 1742-2876
29. https://search.proquest.com/openview/914edbc17cd2ac1b78e214b57461d252/1?pq-origsite=gscholar&cbl=18750&diss=y
30. https://en.wikipedia.org/wiki/Fire_HD
31. https://www.geekwire.com/2018/amazon-passes-samsung-2nd-place-worldwide-tablet-shipments-sees-50-market-share-growth-2017/
32. https://developer.amazon.com/blogs/post/TxDLQFEER6GJ6V/Announcing-the-In-App-Purchasing-API-for-Kindle-Fire-and-Other-Android-Devices.html
33. Chung, Park, Lee, Digital forensic approaches for Amazon Alexa ecosystem. DFRWS 2017 Proceedings of the Seventeenth Annual DFRWS USA. Digital Investigation 22 (2017) S15-S25
34. https://arstechnica.com/gadgets/2018/07/googles-iron-grip-on-android-controlling-open-source-by-any-means-necessary/
35. https://www.emarketer.com/content/amazon-is-now-the-no-3-digital-ad-platform-in-the-us
36. https://forum.xda-developers.com/hd8-hd10/general/discussion-root-progress-fire-hd-8-t3743024/
37. https://developer.amazon.com/blogs/tag/Amazon+Maps
38. https://www.wired.com/insights/2011/09/amazon-silk/
39. https://csrc.nist.gov/projects/mobile-security-and-forensics
40. https://www.howtogeek.com/232973/amazons-fire-os-vs.-googles-android-whats-the-difference/

Mobile Forensics

Cole Troutman and Victor Mancha

Abstract In this chapter, we forensically examine a Samsung Galaxy S7 running Android Version 8.0.0 and an iPhone 7 running iOS version 12.1, and report on the findings of the forensic examination.

1 Introduction

Technology is advancing at an exponential rate in the modern world. With this advancement comes new threats that exploit these new technologies. Therefore, in order to combat these threats, it becomes relevant that an understanding of how these technologies operate be established. In particular, it becomes important to understand how to conduct forensic investigations on these technologies to prevent further attacks from happening. Over the last decade, there have been many changes to both Android and iOS devices. It is vital to continue researching and staying up-to-date on these operating systems and forensic techniques so that one can effectively conduct a forensic investigation for these devices. It is with the goal of understanding how Android and iOS devices operate and how forensic investigations are conducted on these devices that this paper proceeds.

1.1 *Android*

In order to conduct mobile forensics on an Android operating system, it is important to understand the architecture of the Android. The Android has an architecture of layers that stack on top of one another to run. The first layer of the Android architecture is the Applications Layer. This layer focuses on running Native Android applications and Third-Party applications. The second layer of the Android architecture is the

C. Troutman · V. Mancha (✉)
Department of Information System and Cyber Security,
University of Texas at San Antonio, San Antonio, TX, USA
e-mail: Victormancha1@live.com

© Springer Nature Switzerland AG 2020
X. Zhang and K.-K. R. Choo (eds.), *Digital Forensic Education*,
Studies in Big Data 61, https://doi.org/10.1007/978-3-030-23547-5_10

Application Framework Layer. This layer has a series of managers that manage the applications running on the Android. The Application Framework Layer is made up of the Activity Manager, Location Manager, Window Manager, Package Manager, Resource Manager, Content Providers, XMPP Service, and View System [1].

The Activity Manager is responsible for the activity life cycle for all the applications. The Location Manager is responsible for incorporating the Global Positioning System or cellular towers to triangulate the location of the device at all times. The Window Manager is responsible for displaying and hiding all the windows for all the applications. The Package Manager is responsible for installing, uninstalling and upgrading applications. The Resource Manager is responsible for allocating all the proper resources to the applications that are most in need. The Content Providers are responsible for allocating all data being shared amongst the different applications. The XMPP (Extensible Messaging and Presence Protocol) Service is responsible for instant messaging communications and detecting a user's online presence. The View System is responsible for handling all the views for the many different applications [2].

The third layer of the Android architecture is the Libraries Layer. The Libraries Layer stores all the necessary libraries that the Android needs to run a series of applications and programs. The Libraries Layer is made up of SQLite, Free Type, SSL, WebKit, Surface Manager, SGL, OpenGLES, Media Framework, and libc. The SQLite library is a server-less library, meaning that data is pulled and stored from the device without a server. The Free Type library is used for rendering various fonts. The SSL library ensures that all Internet connections are secure. The Surface Manager library is responsible for the composition of windows onto the Android device screen. The SGL library enables the Android device to render 2D graphics. The OpenGLES library is responsible for enabling the Android device to render 3D graphics. The Media Framework library enables the Android device to view images, hear audio, and watch videos. The libc library is the Android systems C library [2]. The Libraries Layer also stores Android Runtime. Android Runtime stores the Core Java Libraries and the Dalvik Virtual Machine. The Core Java Libraries are the libraries that the Android devices use for java applications. The Dalvik Virtual Machine interprets bytecode that was converted from Java byte code into Dalvik bytecode. The Dalvik VM is the key source for Android devices being able to run multiple applications at once [3].

The final layer is the Linux Kernel. The Linux Kernel is the heart and soul of the Android architecture. It is made up of the Display Driver, Process Management, Camera Driver, Power Management, Memory Management, Bluetooth Driver, and Binder Drivers. The Display Driver is responsible for drawing and composing views to be displayed on the Android screen for the user. The Process Management section of the Linux Kernel is responsible for managing different processes that are running concurrently. The Camera Driver is responsible for ensuring the camera of the Android device is functional. The Power Management section of the Linux Kernel is responsible for managing the power of the Android device. The Memory Management section of the Linux Kernel is responsible for managing the memory of the Android device. The Bluetooth Driver is responsible for enabling the Android device to make Bluetooth Connections with other Bluetooth devices. The Binder Driver is used to connect processes to one another and pass parameters and data [3].

Now that a base of understanding has been established for the Android's architecture, it is time to understand how data is stored on an Android device and how this information can be implemented into Android forensics. The first thing to understand is that Android devices store numerous types of data. Data can come from a variety of sources: "SMS, MMS, Chat Messages, Backups, E-mails, Call Logs, Contacts, Pictures, Videos, Browser History, GPS data, Files or documents downloaded, Facebook, Twitter, and other social media application, Calendar appointments, Shopping history, Financial information, Music collection, and Driving directions" [1]. There are plenty of other sources of data that can be collected and analyzed, but the point remains that Android devices can stores large varieties of data. The next thing to understand is where all this data ends up being stored.

Android devices can store data internally and externally. When an Android devices stores data internally, it is usually placed in a predetermined location such as/data/data/<application package>. Key data storage locations are mainly under the "/data/" folder. When an Android devices stores data externally, it is usually to a Secure Digital (SD) card. In this case, a user can move files around and rename folders as they see fit [1]. There are also five methods that can be incorporated to either store or discover data. These methods are: "Shared Preferences, Internal Storage, External Storage, SQLite database, and Network" [1].

The Shared Preferences method incorporates the use of XML files that store key-value pairs in locations such as/data/data/<package_name>/shared_prefs to extract key information. It is through this method that a forensic analyst can extract preferred information about an application such as login names and passwords. Extracting data through the internal storage method is difficult unless the forensic analyst has root access to the Android device. Data stored internally is usually located in the/data/data subdirectory. Depending on the number of applications stored on the Android device, this subdirectory can be quite vast. There is also a structure system located under the applications subdirectory that will include: a cache of files for the application, database files and data, library files required for the application to run properly, and the shared preferences XML files previously mentioned [1]. The external storage method is easier to extract information as there are no firm safe-keeping regulations, such as the requirement of root access. Most Android external storage devices are SD cards in the format of FAT32 FS and store data in a location similar to/sdcard or/mnt/sdcard. The SQLite database method incorporates the use of the APIs Android developers developed so that applications may store data within a SQLite database within the application. The data is usually stored in a location similar to/data/data/<package>/database. The Network method incorporates the use of cloud services that can store Android device data on the network. Some prime examples of these cloud services include Dropbox and OneDrive [1]. It is by these storage methods that data extraction methods can be implemented.

There are two main forms of data extraction for Android devices: logical and physical. Logical extraction focuses on allocated data being extracted from the Android device's file system. This can include capturing images on the screen of the device and software-based logical analysis. Physical extraction focuses on the physical storage medium of an Android device. This can include software-based physical analysis and chip-off analysis, analysis of the Android device's physical chips. It is recommended that all forensic analysts document their steps as they proceed to analyze mobile devices because any change can result in an unwarranted modification on the Android device [1].

To delve a bit further into logical extraction, it is important to understand what type of data can be extracted through this method. Most user data can be extracted through SQLite databases in the Android device such as call logs, text messages, contact information, etc. This data can be extracted using Android Debug Bridge (ADB) but requires for the Android device to be rooted before use. If root access is not possible, an alternative route is to use ADB to backup Android data to a forensic computer for analysis. The only problem with this alternative is that ADB does not backup every application located on the Android device. The most desperate alternative to capturing data on an Android device is to simply take pictures of what appears on the phone. It is a crude forensic method but may sometimes be the only option given to a forensic analyst [1].

Physical extraction methods are deemed harder to extract data, yet the most fruitful because deleted data can be extracted. There are two forms of physical extraction methods: software methods and hardware methods. Software methods focus on providing the forensic analyst with full physical images of data partitions of the Android device. A forensic analyst can use a tool such as FTK Imager or DD to create an image of the Android device, and then use another tool such as Autopsy to perform the actual forensic analysis of the device. Hardware methods focus on physically extracting Android device components to extract physical images. For example, one method of physical extraction is focusing on the JTAG (Joint Test Access Group) interface. The purpose of the JTAG interface is to allow all the pieces of hardware to communicate with each other. By focusing on this piece of hardware, it is possible to recover the flash memory of the Android device and a full physical image. Another physical extraction method that focuses on the hardware revolves around the actual memory chip of the Android device. By heating up the solder until it becomes a liquid, forensic analysts can extract the memory chip from an Android device and recover a full physical image [1]. In the context of this class, only logical extraction methods and software-based physical extraction methods will be implemented for mobile forensic analysis.

1.2   iOS

Before getting into iOS forensics, a brief overview of the architecture of iOS devices is required. Because iOS is a customized version of Mac OS X, there are a lot of similarities between the two architectures and many of the research on the architectural layers came from Mac OS X documentation that Apple provided [4]. iOS has four distinct layers. These layers are, from top to bottom: Cocoa Touch, Media,

Core Services, and Core OS. The Cocoa Touch Layer is the topmost layer and is often referred to as the application layer in iOS. This is the only layer with a slightly different name than that from the Mac OS X layers. This layer provides a number of frameworks that both native applications and developers rely upon.

The Media Layer is directly underneath the Cocoa Touch (Application) Layer. This layer handles the graphics as well as the audio and video [5]. Some of the frameworks for graphics and drawing include, core graphics, core animation, SpriteKit, OpenGL, and GLKit [4].

The Core Services Layer is directly underneath the Media Layer. This layer provides many frameworks that the Cocoa Touch Layer relies on. Some of these frameworks include core data, core location, foundation, and address book. There are many more, but these are some of the most important when it comes to iOS forensics. Many of these frameworks make use of SQLite databases. Even core data, which is often mistaken as a database for iOS developers is more of an object graph manager that just uses SQLite for persistence [6]. The understanding and ability to view SQLite databases will be important in an iOS forensics investigation due to the heavy use of SQLite databases throughout applications and frameworks.

The last layer is the Core OS Layer and is responsible for features related to app security. This layer provides most of the functionality to the above layers. The Core OS Layer encapsulates the kernel environment as well as the low level UNIX interfaces disallowing applications access to what is inside this encapsulation for security reasons [5]. It is this encapsulation at the Core OS layer that makes iOS forensics slightly complicated at times. This is the layer is responsible for the "locked down" feature of iOS. It provides an application sandbox that limits the app's interaction with the system to only what it needs. It also provides a "gatekeeper" that blocks any installation of software that does not come from the app store, which can only be put there by identified developers [4].

Since iOS devices do not have external memory such as an SD card, all data is stored internally (flash memory). When a backup of an iOS device is created using iTunes, you can find the backup data stored in the/var directory. Here you can find user data, installed applications, address book, call history, SMS, Calendar events, Safari Bookmarks and more. Each of examples can be found at a path similar to/var/mobile/Library/<application>. Inside these directories, a couple of the ways that iOS application data is stored can be seen. There are SQLite databases, which many applications heavily rely on. There are also property lists (plists) that can contain basic configurations. Because so many iOS developers have used SQLite (including Apple), a forensic examiner will most likely need a good database viewer at some point in the forensic investigation [7]. Plists are typically in XML format, although a forensic examiner may still see a deprecated NexSTEP or binary format for these file. Plists can be used to store a number of different types of data. These can be strings, dates, Boolean values, numbers [7].

Now that a basic understanding of the architecture and where data is stored, it is time to discuss acquisition methods. There are a few different types of acquisition methods that can be used to conduct iOS device forensics. The ones that will be discussed here are acquisitions via backups and jailbreaking.

Backups are probably the easiest way to conduct forensic investigation on iOS devices. iOS devices can be backed up using iTunes. There are many different tools out there to aid in backup examination. iPhoneAnalyzer by Crypticbit is a good choice for this type of investigation. However, if the backup is encrypted, then other forensic tools may be needed. It is important to note that the evidence collected is forensically sound due to this being executed against the iTunes backup [7]. From here you can easily see the applications and their different folders as well as a manifest.plist, info.plist, Manifest.dp, and status.plist.

Because iOS devices are "locked down" for security purposes, it can make basic forensic techniques slightly more difficult. To get around this, one can Jailbreak the device that is to be investigated. This will allow the investigator to install an SSH server on the iOS device and then from a forensic workstation they would be able to run a dd command to create an image of the iOS device on the forensic workstation [7]. It is important to document any changes you are making to the device so that you maintain the integrity of the image so that it can be used as evidence. For later iOS devices, this could produce an encrypted image that would need the help of other tools such as Lantern Imager or iXAM [7]. There are many options out there to help in jailbreaking a device. Redsn0w and Pangu are good software options to perform the jailbreaking process. After this is done the Cydia application can be installed which is pretty much the app store for jailbroken devices. From here the SSH server previously mentioned can be downloaded and the investigator can extract the image onto the forensic workstation using dd. Once the image is on the forensic workstation, the investigator can use any other forensic tools that they would normally use such as Autopsy or EnCase.

2 Literature Review

2.1 Android

According to Hazra and Mateti [8], Android forensics can be classified into two categories: Proactive and Reactive forensics. Proactive forensics depends on an investigator preparing for an event to occur in order to capture data in real time. This could be capturing text messages, phone calls, video chat sessions, etc. Reactive forensics are more familiar and known as traditional forensic techniques. This means that once a mobile device is confiscated, it is placed under the usual data extraction techniques, such as imaging, carving, etc. The difference between the two depends not only on separate technologies, but law requirements as well. It is through this distinction that Android forensic techniques advance with technology [8].

Mylonas [9] discusses his team's approach towards proactive forensics. The technique is dependent upon three separate entities. The first entity would be considered the investigator. The second entity would be considered an independent authority who handles data collection and storage. The third entity would be the suspect that the investigation is being conducted on. The proactive forensic technique is highly dependent upon a software agent being placed in the suspect's mobile device in order for it to work. Once all of the prerequisites are met, the technique is executed in a set of six stages. The first stage requires the investigator to submit an investigation request and wait for approval. Once it has been approved, the investigator must identify the type of evidence they are trying to acquire. If it gets approved, the software agent will be free to begin collecting evidence. Once evidence has been collected, it then becomes a matter of transferring that evidence back to the independent authority and then storing that evidence in a safe location where it cannot be tampered with. The final stage includes completing the investigation, hopefully with the arrest of a guilty criminal. There are legal issues with this proactive forensic technique that are dependent upon the judicial system for its success, but the results can be described as worthy of waiting [9].

In terms of reactive forensic techniques, there have been many scholars who have proceeded in understanding the unique ways of data extraction for Android mobile devices. The most significant difference between all of these techniques is dependent on if the android is rooted or not. Root access means that the user has administrative privileges and can be seen as dangerous if a malicious force were to gain these privileges. Although it is very difficult, Lucideus Research [10] discusses the different acquisition techniques for Android mobile devices. There are three main variables that determine the level of data acquisition that can be acquired from an android mobile device. The first is the security and version of the device. It would prove extremely difficult to use acquisition techniques if the investigator were unable to even access the device. It would also prove tedious to use techniques that do not properly function on a specified version of the Android device. The second factor is where or not the device has root access. The final factor is the connection medium the investigator is using to perform acquisition techniques on the android. This can be through the use of Bluetooth, Wireless, and/or Wired, which is the preferred method of connection. In the event that the Android is rooted, it is noted as very simple to create an image of the device using forensic tools or manual techniques. The use of Cellebrite UFED, Oxygen Forensics Suite and Magnet Axiom are all highly recommended as easing the process of imaging an Android device. As for the manual process, the tools DD and FTK Imager are recommended for manual imaging. As for non-rooted devices, there only exists the option for logical acquisition through the use of Android Debug Bridge (ADB) and creating a backup of the entire Android device. The backup will allow the investigator to save system and application data, which may prove enough for their investigation [10].

2.2 iOS

Murphy [11] at Gillware Digital Forensics conducted a non-criminal case that showed the importance of iOS forensics using backups. A husband and his wife were tragically killed in a car crash in February of 2017. The couple had two kids. The parents of the husband brought his phone into Gillware to try to salvage some of the data pertaining to the children. There was no passcode on the device, and the credentials were known for the iCloud account. However, the backup that was on iCloud was encrypted. Murphy admitted that though there were methods to break the encryption through use of brute force tools available, it is an extremely time-consuming process that should be weighed against other options. At the time, iOS 11 had just been released and with it came an option that was not previously there before, which was the option to remove the encryption key by resetting the general settings on the device. This would not help gain access to previously made backups, but it would allow to make new backups from the device without encryption. The device in question was running iOS 10.3. With the parent's consent, they were able to update the son's device to iOS 11 in order to make an unencrypted backup and extract the desired data and artifacts successfully [11].

Katalov [12] of ElcomSoft Desktop Mobile & Cloud Forensics wrote an article about iOS forensics from iOS 4 up to iOS 11. He discusses some of the different methods to conducting iOS forensics as well as the pros and cons of each. One method is the logical acquisition via backups. This method is the fastest and easiest way to obtain data from an iPhone. The other method is the physical acquisition. For this method a jailbreak is needed in order to set up an SSH client and extract each bit from the device. Though this method may take longer and make some changes to the actual device (which can be a big deal in some forensic cases), it will result in the most information extracted from the device. Some of the data that can only be found from a physical acquisition is the location data, logs of connections to 2G/3G/4G networks, downloaded mail, and more. Katalov also discusses that even though physical acquisition techniques can extract every bit of the device, it might not be as helpful as one would think when it comes to carving for deleted data because, since iOS 4, the actual data is still there, but encrypted. When it is deleted, and the space is unallocated, then the encryption keys are wiped leaving no way to decrypt the file. There are still some pieces of evidence that might remain if deleted and carved such as record data from SQL databases. Katalov briefly talks about cloud acquisition in this article as well. He mentions how Apple designed iCloud backups to only be able to restore a device. For example, one would not be able to just download the backup files directly from the cloud. However, ElcomSoft has developed software to be able to obtain these backup files without having the device present by using a Man-in-the-Middle technique. This allowed them to be able to "capture the traffic during the restore process, analyze the protocol, and reverse the encryption" [12].

The SANS Institute Reading Room posted an article by Proffitt [7] which discussed iOS forensics. Once again there are the main methods of acquisition by iCloud, backups from iTunes, and physical extraction through jailbreaking. This article went more in depth on how items are stored on the device and what the forensic investigator should be looking for. Much of the data found in an iOS forensic investigation will be in the form of a plist. This is a property list to store various types of data on iOS and Mac operating systems and usually in XML or binary format. There is also SQL databases that a lot of applications use to store bits of data as well. When a forensic investigator finds timestamps for MACB times, it will be in CF Absolute

Time. Proffitt wrote that applications will usually have .db or .plist files where the investigator can find usernames, sometimes passwords, cookies, or images that can help provide evidence [13]. This article also provides many helpful ideas for where to find artifacts during the iOS forensic investigation. This included the artifacts: Keyboard log of keystrokes (to help in predictive text) in a .dat file, Notes that are found in Library/Notes in a .sqlite file, text messages in a sms.db file, address book in address.sqlite file, and more [7].

Infante [14] wrote a blog post that discusses his look into the iOS backup file, what its contents are, and how to navigate it. A lot of this information is his own notes to himself, so as he states to everyone reading it, it is best not to consider this fact, but rather as a base for their own research. Some of the items of interest are the info.plist, manifest.plist, manifest.db, and status.plist files. These are included in every backup file and contain the information about the device and backup files. The info.plist is the file that contains information about the device such as serial number and even telephone number if applicable. The manifest.plist and status.plist files contain information about the backup. The manifest.db is the actual database that holds records of all the files stored in the backup. This is the file that will be of most interest to forensic investigators. Included with these information files are the actual backup files. The files are named in a possible SHA-1 format and sorted into folders beginning with the first two characters of the filename [14].

3 Identification of Research Question

There have been many changes to the operating systems of both Android and iOS mobile devices since they first were created. Some of these changes are obvious like the user interface. However, there are more subtle changes from version to version that a forensic examiner must have up-to-date knowledge on. Though it would be helpful to understand all these changes and how each version of a particular device differs from that of the other versions, one question that will always be important in mobile forensics is: How are forensic investigations conducted on current Android and iOS devices?

In order to conduct more comprehensive research on these methods this semester, the scope of this research will narrow to include only a small selection of these methods. For Android devices, only logical and software-based physical extraction methods will be analyzed. For iOS devices, only logical acquisition methods via backups will be examined.

4   Proposed Research Methodology

4.1   Android

In order to properly conduct an investigation on an Android mobile device, there are a number of variables that must be determined. As previously stated, there are three key prerequisites: the device's security and version, the device's accessibility, and the device's connection medium. The most current Android version is 8.0.0 and the investigator should be able to bypass the security if a proper investigation is to be conducted. Both logical extraction through the use of a non-rooted Android device and software-based physical extraction through the use of a rooted Android device will be explored in this research. The connection medium for both of these forensic techniques will be USB connectivity with the MTP connection mode being enabled. This will allow for the internal memory of the device to be logically accessed. The USB debugging mode will also be enabled to allow for the Android Debug Bridge to send commands to the Android device [10]. There should also be a proper forensic workstation that stores the necessary tools to conduct the investigation. This workstation should include the Android Debug Bridge (ADB), an imaging creation tool such as DD or Cellebrite UFED and an image reader such as FTK Imager or SQLite Database Browser. The workstation can be either Windows or Linux, but will be Windows proceeding forward in this research.

Once the workstation has installed all of the necessary tools, the Android forensic techniques can commence. The first step to conducting a logical extraction of a non-rooted Android device is to ensure all of the prerequisites are met (updated software, bypassable security, non-root, and USB connectivity with MTP and USB Debugging enabled). The investigator should connect the Android device to the workstation and expect a message that asks the investigator if the Android device should trust the workstation. Once the investigator selects yes, the next step would be to open up command prompt and locate the ADB executable. With the Android device connected, the investigator should ensure ADB finds the device through the command "adb devices". This will provide the user with a list of devices that are attached and a status of "device". When the Android device is confirmed to be found, the next step would be to create a folder that will store the backup. For all intensive research purposes, the folder can be named something simple, like "image" and stored in the same location as the ADB executable. The user must change the directory to this new folder and execute the command "adb backup-all-f backup.ab". This will then open a prompt on the Android device that will ask the investigator if they want to conduct a full backup. The investigator has the option to encrypt the backup and proceed. Once this is done, the investigator will have complete access to the logical extraction of data for the Android device. Through the use of a JAR file called abe (android backup extraction), the investigator can unpack the backup into a tar file. The tar file can then be read by a tool such as FTK Imager. The tool FTK Imager can then be used to analyze and extract further data. For example, in the event that the user had a German application called Pixel House, the backup would have a database known as chefkoch_db. This file can be extracted and opened in a database reading tool such as DB Browser for SQLite. This tool will then inform the investigator of the Tables, the Indices, the Views and the Triggers of the database [10]. The Android backup tar file will provide the investigator with numerous helpful databases, such as Calendars and user preferred Bookmarks, but it will keep many other important folders hidden such as Contacts and Browser data. The reason behind this is due to privileges and security. In the event that the investigator uses a rooted Android device, the backup will provide the investigator with many more databases that can be utilized for an investigation.

The process of conducting a forensic investigation on a rooted Android device is a bit lengthier. The beginning prerequisites are all the same (updated software, bypassable security, and USB connectivity with MTP and USB Debugging enabled). However, there is one big difference between the two and that is root privilege access. The process of rooting an Android phone is lengthy, arduous and risky. For the sake of this proposed research methodology, it will be assumed that the phone has not been rooted. In this case, the investigator must understand that different versions of Android will lead to different methodologies of rooting the device. In the past, a methodology involving BusyBox and KingoRoot would have sufficed the criteria in a matter of minutes. Unfortunately, as time progressed, so did the way Samsung prevented root access to its consumers. In today's time, the Android 8.0.0 can be rooted through the use of a number of tools: TWRP (Team Win Recovery Project) file, Odin, No Verity and Magisk. Once all of the tools have been installed on the workstation, the process involves the user to copy No Verity and Magisk onto the SD card of the Android device. The user must then unplug and turn off the Android device and start it in download mode. On the Samsung Galaxy S7, this is achieved by holder the power button, volume down button and home button for about 5–10 s. The user must press the volume up button to continue, replug the device into the workstation and start Odin. Through Odin, the user must select "AP" and the downloaded TWRP file and select "Start". Once the file is complete, the user must turn off the Android device and restart it in recovery mode by pressing the power button, the volume up button and the home button. When the device starts in recovery mode, the user must select "Wipe", "Advanced Wipe", "Data", "Repair or Change File System", "Change File System", "EXT2", "back", "EXT4", back to the opening screen, "Reboot", "Recovery", "Do not install", "Install", "Select Storage", "Micro SD Card", "OK", "No Verity", "Back", "Magisk", "Reboot System", "Do Not Install", "Select APN" and "O2 UK Pay & Go" [15]. Afterwards, the investigator will have to set up the phone due to its reinitialization, but the Android device will be fully rooted. Although this was the only successful method to root an Android phone that the researcher found, it completely misses the mark by wiping all of the users data on the device. Therefore, proceeding forward in this proposed research methodology, it will be assumed that the investigator found an alternative to rooting the mobile device without deleting all of its content. Once the device has been rooted, the investigator can create an image from the Android device's partitions. Through the use of FTK Imager, or a similar tool, the investigator can inspect the newly created image and proceed forward with their investigation. The previously missing

Contacts and other assorted hidden folders will become available to the investigator. The database files can then be retrieved and the use of a database reader, such as DB Browser for SQLite, can be utilized to extract all of the necessary data for the investigator's investigation [10].

4.2 iOS

For a mobile forensics investigator, it is important to understand how to conduct these investigations on iOS devices using the logical acquisition through backups method. This method has always been, and still is, the fastest and quickest way to gather up much of the data from iOS devices [12]. Even though a logical acquisition does not give you the full bit-by-bit image, it is still important as seen in the case that Murphy [11] conducted. Also, there is not too much of a difference from backup forensics and full physical extraction methods since it will be very difficult if not impossible to carve for many deleted artifacts in the unallocated space of a physical extraction since the encryption key is wiped when the space is unallocated [12].

In order to perform research on this forensic technique, one must have access to an iOS device. In keeping with the main research question of "current" devices, this iOS device should not be running on anything earlier than iOS 11 as currently the latest version is iOS 12.1. The next thing that one needs is a forensic workstation. This is the area that will the forensic investigator will keep all of their tools to aid in the investigation. For the purposes of this research, it may be better if a workstation running Mac OS was used. As discussed in the introduction/background information on iOS devices, much of the data that will be found on iOS devices are stored in plists and sqlite databases. There are tools out there such as iPhone Analyzer or iMazing that layout a backup file in an easy to view way. Tools like this may be helpful in an actual investigation, but for research purposes, it might be more beneficial to find these files manually in order to gain a better understanding of the iOS device and how it works. In order to do this, a few tools will still be needed in order to view some of these files. The developer tool, Xcode, should allow the forensic examiner to view any plist file. This is a free tool that can be downloaded from the Mac Application store. The next tool that will be needed is a sqlite database viewer. There are many free choices on the internet such as DB Browser for SQLite. The last tools that one might need for an investigation will be iTunes (for obtaining the backup file) and a simple text editor for reading text files. These are both already installed on Mac OS systems.

Once these tools are downloaded and installed and the forensic workstation is all ready to go, the process of producing a backup can begin. As previously mentioned, the easiest method for this requires iTunes. From here the backup can be made. For the purposes of this research, it is easiest if the backup is not encrypted. There are methods for breaking encryption, but as they are time consuming and not the purpose of this research, this will be skipped. Once the backup is made, the investigator may find it necessary to obtain the md5 hash of this file in order to be able to prove that no changes were made during the investigation. However, for the purposes of this research again, this can be skipped. Once the files are located and copied to the forensic workstation, the analysis can begin.

The first thing that the investigator should see are the four files; info.plist, manifest.plist, manifest.db, and status.plist. The investigator should be able to open these plist files with xcode to find any general information about the device as well as information about the backup itself. The next thing the investigator should do is to open up the manifest.db file using the SQLite database viewer and then the process of looking for artifacts can begin.

There are many different types of artifacts that might be of interest to the forensic investigator. Using the database viewer, the investigator can browse through the files and entries for common artifacts. From here the investigator will be able to copy and paste that string into the Finder on Mac and be able to find that specific file from the backup folder. Once the file is found, the investigator can copy and paste the file into a new folder of artifacts of interest. This process can be repeated for each artifact that the user wants to find.

Another type of data that the investigator can look into is application data. Sometimes this data will not be stored locally on the device and therefore will not show up in the backup. However, there are some apps that may cache bits of information on the user's device so that it loads quicker during startup and saves the user from having to download it each time. Some of these types of items can be images or saved messages particularly from applications with chat history. One way that might be beneficial for one researching and trying to know what type of data they can try to look for on a device would be to switch their iOS device into airplane mode and then browse some of the apps that may or may not store data on the device. A quick look at Instagram while the device was in airplane mode still contained a few messages from each thread that was cached and stored on the device. This method will give the investigator a possible target to look for when browsing through the manifest database file.

Once the investigator has finished exploring the manifest.db file and has collected all of the artifacts of interest, then they can explore the artifacts and see what information they can pull from them. This part may be a bit tricky because different applications store data a little bit differently. Some knowledge of how SQLite databases work (such as knowledge of primary keys and how they relate to different tables in a database) will be of benefit to the investigator. Even some knowledge of SQL queries might help to extract data faster. But even without this knowledge, it is possible to browse through the databases and find specific records that might be of interest.

This would conclude the method of a manual forensic investigation of iOS devices via backups. From here the investigator may wish to explore some automated tools that aid in browsing backup files such as the ones previously mentioned. They may also wish to look into other methods of acquisition such as jailbreaking the iOS device and installing an SSH server in order to remotely log in and extract a physical image using the dd command or similar.

5   Experimental Results and Discussion

5.1   Android

For the experimental research into Android forensics, the two approaches previously mentioned were utilized. The Android device used for the experimental research was a Samsung Galaxy S7 running Android Version 8.0.0. The forensic workstation was a 2011 iBuyPower gaming laptop installed with Android Debug Bridge (ADB), Android Backup Extraction (ABE), FTK Imager, DB Browser for SQLite, and various versions of Odin, TWRP, ncat, CF Auto Root, and Samsung Galaxy S7 drivers for Windows.

The first step towards the logical extraction of data from a non-rooted Android device required that the device be updated to Android version 8.0.0, the security was bypassable, the connection was MTP and that the USB debugging mode was enabled (Fig. 1).

The next step required the forensic investigator to ensure that adb could find the device (Fig. 2) through the use of the command "adb devices".

Once the device was found, the investigator created a new folder called "image". The directory was changed to this new folder and the command "adb backup -all -f backup.ab" was utilized to create a new backup of the Android device (Fig. 3).

Fig. 1 Android prerequisites

```
C:\Users\user\Desktop\SDK_Tools\platform-tools>adb devices
List of devices attached
* daemon not running; starting now at tcp:5037
* daemon started successfully

C:\Users\user\Desktop\SDK_Tools\platform-tools>adb devices
List of devices attached
f2c947f4                device
```

Fig. 2 Android device found

```
C:\Users\user\Desktop\SDK_Tools\platform-tools>cd image

C:\Users\user\Desktop\SDK_Tools\platform-tools\image>adb backup -all -f backup.ab
Now unlock your device and confirm the backup operation...
```

Fig. 3 Creation of android device backup

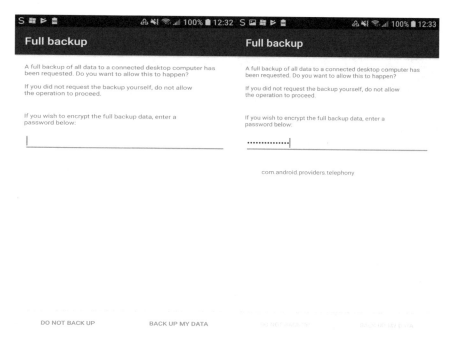

Fig. 4 Agree to the backup

The Android device issued a message asking the user if they wanted to perform a backup of all the data on the device. The investigator agreed and inserted a password (Fig. 4).

The investigator proceeded by using the abe.jar to unpack the newly created backup. Due to the fact that a password encrypted the backup just moments before, the same password had to be utilized to decrypt all of the data (Fig. 5).

```
C:\Users\user\Desktop\SDK_Tools\platform-tools\image>java -jar abe.jar unpack backup.ab backp.tar ""
Backup encrypted, enter password (will NOT be displayed):
Password:
```

Fig. 5 Unpacking the backup into a TAR file

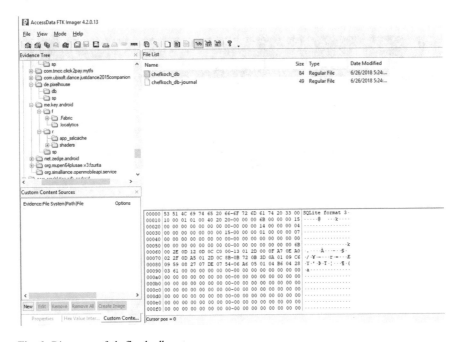

Fig. 6 Discovery of chefkoch_db

With the newly created TAR file, the investigator imported the image into FTK Imager. It was here that the investigator found numerous interesting items such as chefkoch_db, calender.db, partnerBookmarks.db and much more (Fig. 6).

The investigator proceeded to extract the databases and import them into DB Browser for SQLite so that they could be properly read. The chefkoch_db displayed 5 German recipes, the ratings, and the Booleans for images and videos (Fig. 7).

The Calendar database revealed 3 separate email accounts for the user of the device: "v.kramer1990@gmail.com", "veronika-k@gmx.net", and "htc895@my.utsa.edu" (Fig. 8). This information could prove very useful in an investigation that is dependent upon discovering a user's email addresses and personal calendar events.

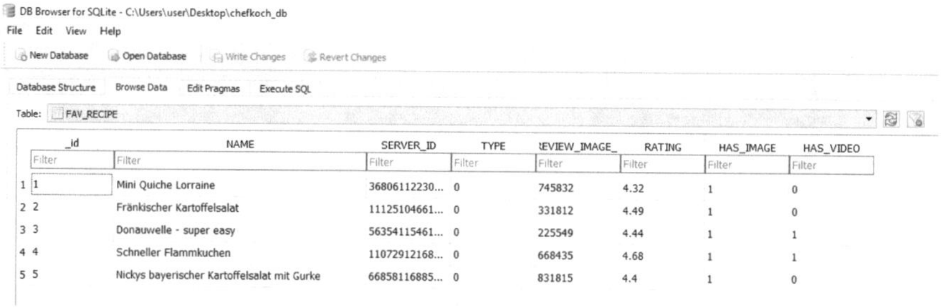

	_id	NAME	SERVER_ID	TYPE	ιEVIEW_IMAGE_	RATING	HAS_IMAGE	HAS_VIDEO
1	1	Mini Quiche Lorraine	36806112230...	0	745832	4.32	1	0
2	2	Fränkischer Kartoffelsalat	11125104661...	0	331812	4.49	1	0
3	3	Donauwelle - super easy	56354115461...	0	225549	4.44	1	1
4	4	Schneller Flammkuchen	11072912168...	0	668435	4.68	1	1
5	5	Nickys bayerischer Kartoffelsalat mit Gurke	66858116885...	0	831815	4.4	1	0

Fig. 7 Chefkoch_db in DB browser for SQLite

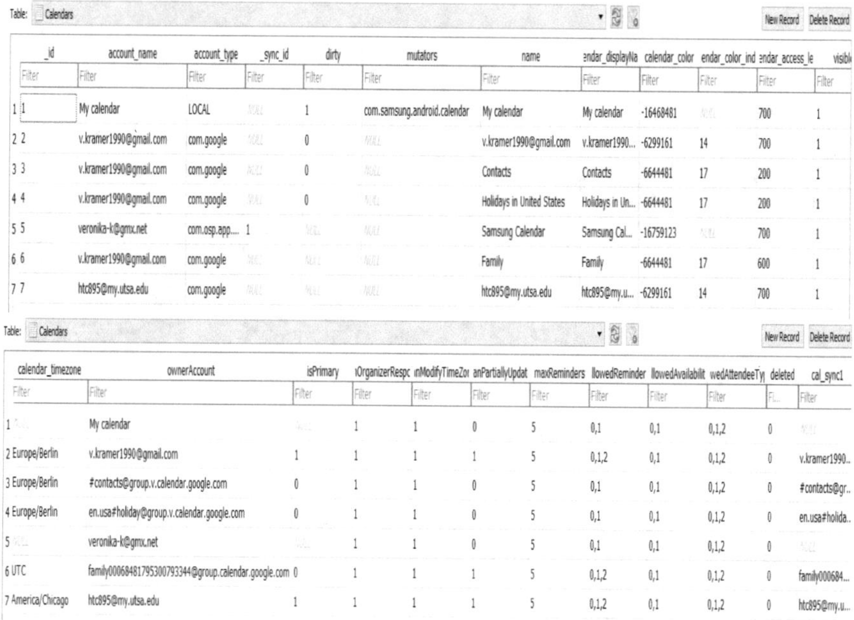

	_id	account_name	account_type	_sync_id	dirty	mutators	name	ɜndar_displayNa	calendar_color	endar_color_ind	ɜndar_access_le	visibl
1	1	My calendar	LOCAL	NULL	1	com.samsung.android.calendar	My calendar	My calendar	-16468481	NULL	700	1
2	2	v.kramer1990@gmail.com	com.google	NULL	0	NULL	v.kramer1990@gmail.com	v.kramer1990...	-6299161	14	700	1
3	3	v.kramer1990@gmail.com	com.google	NULL	0	NULL	Contacts	Contacts	-6644481	17	200	1
4	4	v.kramer1990@gmail.com	com.google	NULL	0	NULL	Holidays in United States	Holidays in Un...	-6644481	17	200	1
5	5	veronika-k@gmx.net	com.osp.app....	1	NULL	NULL	Samsung Calendar	Samsung Cal...	-16759123	NULL	700	1
6	6	v.kramer1990@gmail.com	com.google	NULL	NULL	NULL	Family	Family	-6644481	17	600	1
7	7	htc895@my.utsa.edu	com.google	NULL	NULL	NULL	htc895@my.utsa.edu	htc895@my.u...	-6299161	14	700	1

	calendar_timezone	ownerAccount	isPrimary	ıOrganizerRespc	ınModifyTimeZoı	anPartiallyUpdat	maxReminders	llowedReminder	llowedAvailabilit	wedAttendeeTy	deleted	cal_sync1
1	NULL	My calendar	NULL	1	1	0	5	0,1	0,1	0,1,2	0	NULL
2	Europe/Berlin	v.kramer1990@gmail.com	1	1	1	1	5	0,1,2	0,1	0,1,2	0	v.kramer1990..
3	Europe/Berlin	#contacts@group.v.calendar.google.com	0	1	1	0	5	0,1	0,1	0,1,2	0	#contacts@gr..
4	Europe/Berlin	en.usa#holiday@group.v.calendar.google.com	0	1	1	0	5	0,1	0,1	0,1,2	0	en.usa#holida..
5	NULL	veronika-k@gmx.net	NULL	1	1	0	5	0,1	0,1	0,1,2	0	NULL
6	UTC	family00068481795300793344@group.calendar.google.com	0	1	1	1	5	0,1,2	0,1	0,1,2	0	family000684...
7	America/Chicago	htc895@my.utsa.edu	1	1	1	1	5	0,1,2	0,1	0,1,2	0	htc895@my.u...

Fig. 8 Calendar.db evidence

Although it was not very informative, the personal Bookmarks database revealed bookmarks could be retrieved from a backup of an Android device and could potentially be helpful on an investigation of a user who highly utilizes them (Fig. 9).

The remaining databases of the Android backup displayed similar information that was not quite useful due to the nature of this research project. In the event that a specific goal was in mind, for example, exclusive retrieval of all SMS messages and Call Logs, then it might have been more helpful to extract this data and focus all attention on this goal. Otherwise, this logical extraction of data on a non-rooted Android device was a success.

_id	title	url	type	parent	favicon	touchicon
Filter	Filter	Filter	Filter	Filter	Filter	Filter
1 1	Samsung Mob...	NULL	2	0	NULL	NULL
2 2	AT&T Mobile ...	http://home.att.com	1	1	NULL	NULL
3 3	AT&T Wi-Fi H...	http://attwifi.knowwhere.com/attwifiw/	1	1	NULL	NULL
4 4	AT&T Reuse ...	http://www.att.com/recycle	1	1	NULL	NULL
5 5	Device Help	http://www.att.com/devicehowto/?make=samsung&model=samsungg930a	1	1	NULL	NULL

Fig. 9 Personal bookmarks

```
C:\Windows\System32\cmd.exe

Microsoft Windows [Version 10.0.17134.407]
(c) 2018 Microsoft Corporation. All rights reserved.

C:\Users\user\Desktop\SDK_Tools\platform-tools>adb devices
List of devices attached
* daemon not running; starting now at tcp:5037
* daemon started successfully

C:\Users\user\Desktop\SDK_Tools\platform-tools>adb devices
List of devices attached
355a384c344f3098          device

C:\Users\user\Desktop\SDK_Tools\platform-tools>adb -d install KingoRoot.apk
Success

C:\Users\user\Desktop\SDK_Tools\platform-tools>adb -d install BusyBox.apk
Success

C:\Users\user\Desktop\SDK_Tools\platform-tools>
```

Fig. 10 Successful installations of KingoRoot.apk and BusyBox.apk

As for the physical extraction of data from a rooted Android device, the experimental results did not align with the proposed research methodology. The investigator did their best to perform the specified steps as precise as possible, but to no avail. The investigator began the experiment with very little knowledge about the process of rooting and originally tried utilizing the BusyBox and KingoRoot approach. The idea was simple, ADB would install two APKs into the device and root the device in a matter of minutes. Unfortunately, as the first part was successful, the second half was not (Fig. 10).

KingoRoot failed to root the Android device due to its inability to root any Android with a version past 5.0.0. This created a problem seeing as the experimental Android had a version of 8.0.0. The next step became a big process of trial and error as indicated by the assortment of tools downloaded into the workspace to try to root the Android device (Fig. 11). None of which led to a successful end. It was not until the finding of Android Doctor's tutorial that a working understanding of Android rooting was established. Unfortunately, Android Doctor's tutorial only covered International phones. The process remained the same, but the United States version of the TWRP file was impossible for the investigator to find for the specific Android device that

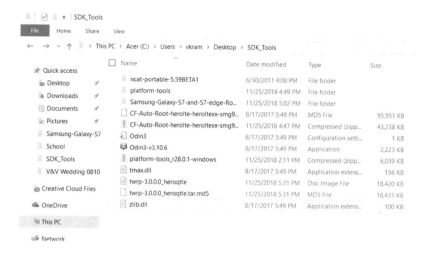

Fig. 11 Android root tools

was being used. Although unsuccessful, the investigator understands that once the device was rooted, it would have been a simple matter of utilizing a tool such as DD to create an image of the device.

Once an image was created, the next steps would have been very similar to the first logical data extraction approach. The use of FTK Imager would have been utilized to extract data such as Contacts, Pictures and Browser History. Then through the use of DB Browser for SQLite, the results would have been analyzed and processed for further investigation dependent upon the end result goal. With that, the experiment of Android forensics data extraction was concluded. In the end, it proved more reliable to use the non-root, logical data extraction method to investigate an Android device. Although the root method would have provided the investigator with more knowledge about the Android device, the results suffice for the purpose of this research.

5.2 iOS

For the experimental research into iOS forensics via backups, the same approach that was proposed earlier was used. The device used for the experimental research was an iPhone 7 running iOS version 12.1. The forensic workstation was a 2012 MacBook Pro running Mojave version 10.14.1. Xcode had already been downloaded on the Mac, however there was no SQLite database viewer. DB Browser for SQLite is a free database viewer that was installed for the purposes of this project and it ended up worked very well for this research. iTunes and TextEdit (a basic text editor) were already installed on the Mac. With the forensic workstation set up, it was time to dive into the forensic experiment.

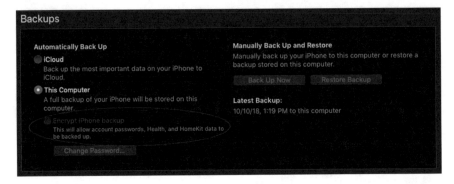

Fig. 12 Encrypt iPhone backup un-checked

The first thing that was done was to plug the phone up to the Mac and open iTunes. When navigating to the backup controls on iTunes and before starting the backup process. The "encrypt this backup" checkbox was unchecked so as to make the research a bit easier. This is shown in Fig. 12.

After the device was finished being backed up to the Mac workstation, the rest of the research could continue, starting with finding the location of the backup file. This was found on the Mac at ~/Library/Application Support/MobileSync/Backup. No other backup had previously been made on this Mac, so the only file there was the backup folder that was to be investigated. If there were other backups present, then it would be easy to find the backup by the time in the Finder window, or by viewing the info.plist file and checking the last backup date as shown in Fig. 13. Note that the Mac will not allow you to open or view any file in this location. In order to view these files one must copy the backup folder and paste it into a different location such as in documents or on the desktop where they can then begin to open files and analyze the backup.

Looking more into the Info.plist will show a lot more information about the device itself such as the phone number, the product name, the version of iOS the device is running. Also it is important to note that any of the "Data" types contain base64 encoded data that is usually another plist. However, for this research, these data values were not looked into at depth. After looking at the Info.plist, it was time to look at the manifest and status plists. These gave information about the manifest database and information on the backup itself respectively. Then the database file, manifest.db, was opened using DB Viewer for SQLite. This is shown in Fig. 14.

This database viewer was very helpful in finding artifacts of interest mainly because it allowed the user to type in filters at the top of each column to search for instead of having to execute SQL queries even though there was an option for that as seen in the right tab in Fig. 14. The leftmost column is the fileID, this is the actual filename that can be located in the backup file. The Domain column shows under which domain the file falls under. Then it gives the path location in the rela-tivePath column. This column was found very useful for searching for the types of

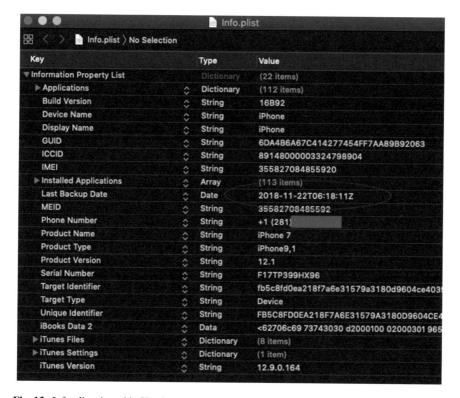

Fig. 13 Info.plist viewed in Xcode

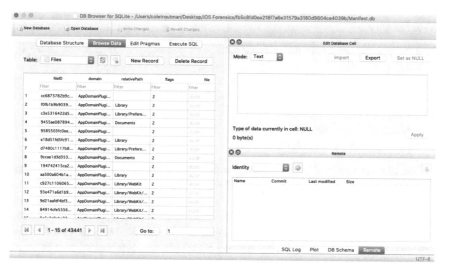

Fig. 14 Manifest.db files table view

file that would yield helpful data and artifacts. For example, if one typed in .db, then they would get a list of all database files in the backup. You could narrow this search down to typing in a certain domain as well which really made zeroing in on certain artifacts a breeze if you did not quite know the full name or path of it. The other two remaining columns were not used, though it is helpful to note that when "BLOB" is seen in an entry, it refers to binary plist that seems to be metadata on the file entry. This was not confirmed though in this research.

From here, it was time to start looking for certain artifacts. The first artifact that was found was the CallHistory.storedata. An easy way to locate this file is to type in "CallHistory.storedata" in the relativePath column and it should be the only entry with that name. Then copy and paste the fileID into a Finder window search bar and the file will pop up soon. This file can then be copied and placed into an "Artifacts" folder or similar. From there the researcher/investigator can rename the file to something easier to read (preferably "CallHistory.storedata") and if the extension is something recognized by the Mac, then it will open it up in the default application. If the default application was not the application that you wanted to open the file in, then right clicking the file and specifying which application to open it with will always work. From there the CallHistory.Storedata database was able to be viewed in previously mentioned database viewer. This is shown in Fig. 15.

From this image, one can see that a lot of data is shown, and even more data can be found throughout the multiple tables in the dropdown menu. This current table shows a cropped entry selection that reveals all the calls that were made or received on the device as well as location. On the right you can briefly see the first 3 digits of the telephone numbers under the ZADDRESS column. If a forensic investigator were to try to piece together the caller to the number, they may be able to find more information in the Address Book file. This was the "formula" for finding the rest of the artifacts:

1. Type the filename into the relative path
2. Copy the fileID
3. Paste it into a Finder search bar
4. Copy the file that pops up
5. Paste it into Artifacts Folder
6. Rename file with proper extension
7. Open file and view contents

This can be repeated over and over until enough artifacts are found to appease investigator's curiosity. The results of all the artifacts found during this research is shown in Fig. 16.

A brief description of some of the artifacts found will be given in this section. In the CameraRoll folder, the Photos.sqlite database is a database file that contains information on every photo in the camera roll. This was not looked into as much because it would be better to have an automated tool find and extract the hundreds of photos on this iOS device. SMSHistory.db contained a database of every single message that the device contains. The messages were sorted chronologically from when they were received and the sender phone number was actually in a different

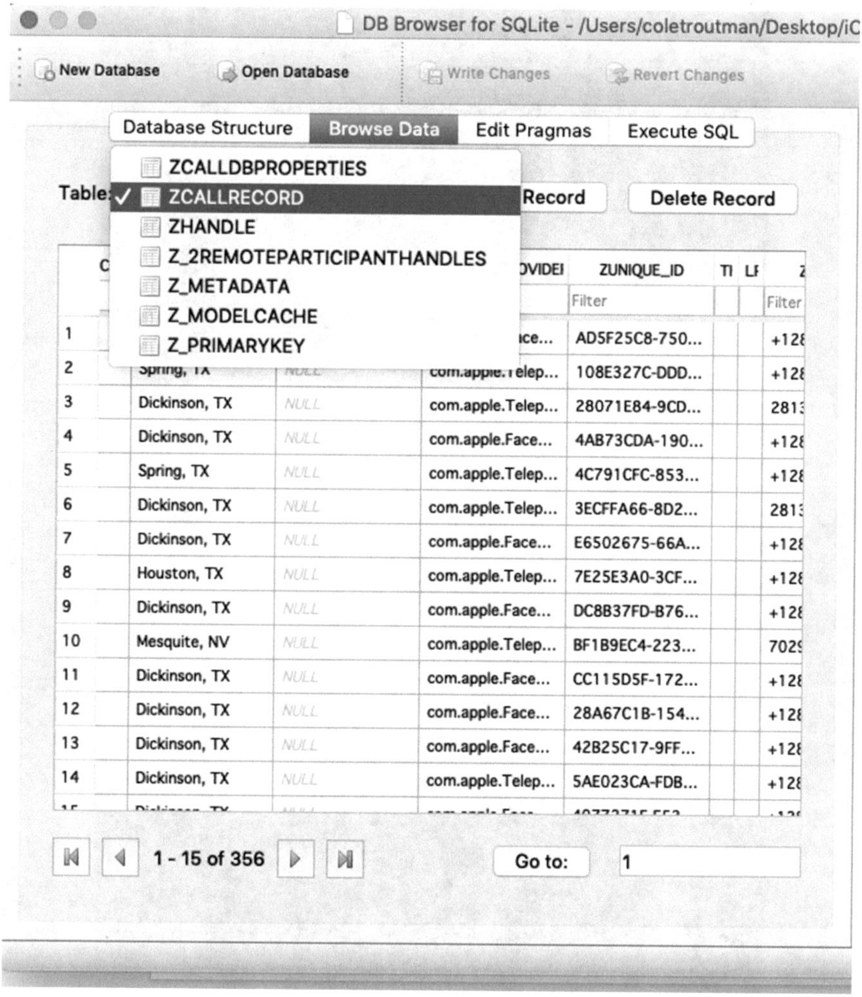

Fig. 15 CallHistory.storedata ZCALLRECORD table view

table than all the messages and required the investigator to take the handleID from the messages table, and search for it in the handle table. Again this resulted in just the phone numbers (emails for some iMessages) that could then be looked up in the Address Book database on the phone if desired. All the notes stored on the phone were found in the notes.sqlite file. The safari History was all shown in the SafariHistory.db file. One of the more interesting artifacts that was found was the Instagram artifacts. This was interesting because after switching the device to airplane mode and searching around social media apps for any data that might be stored, it turned out that Instagram caches the last 10 messages per thread on the device whereas FaceBook must have stored all the messages on their server because

Fig. 16 Finder view of artifacts found

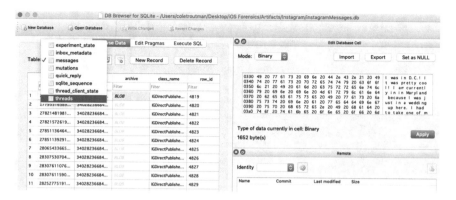

Fig. 17 Instagram DB

none of those messages showed up when in airplane mode. From the Info.plist, the unique bundle identifier was found for the instagram application. This was then used to filter for the domain column in Manifest.db. In the relative path a ".db" or ".sqlite" filter was applied and then the database for instagram messages was found. After extracting this artifact and opening it up the result is seen in Fig. 17.

In this image one can see that list of tables in the dropdown menu. It was discovered that the threads and messages were the most important for an investigator. Off to the right in the database cell the actual message is seen in a binary plist format. To obtain this data, all one has to do is click on the "Export" button in the upper right and choose the location and file name for that data. From there one can view the file in Xcode and see the information and message more clearly. In the threads table, one is actually able to link the Instagram username of the sender and participants of each thread. From there one can obtain the thread id and go back to this messages table in Fig. 17 to filter for that ID and find all the messages from that thread. This was very interesting to see how Instagram stores some data on the user's device. For the purposes of this research, a full conversation was pulled between Cole and Ruthy talking about her new baby (as shown in all the plists under the Instagram folder in Fig. 16) and was compiled in an Excel Document as shown in Fig. 18.

The NSTime (CF Absolute Time) could be converted to a human readable time but this was enough for the purpose to sort the messages in ascending order by NSTime to get the chronological timeline of each message.

From there the experiment into iOS forensics via backup was concluded. Further research, as previously mentioned, would delve into physical extraction via jailbreaking the device to see if any deleted data in unallocated space (that could not be seen through logical backups acquisition) could be seen and extracted.

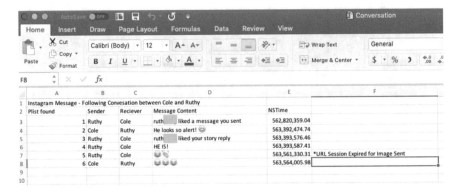

Fig. 18 Conversation found in instagram

References

1. Santos, N.: Mobile Forensics: Android. Retrieved 6 Oct 2018, from https://fenix.tecnico. ulisboa.pt/downloadFile/1970943312266679/csf-12.pdf (2015)
2. Sojitra, M.: Android Application Framework. Retrieved 6 Oct 2018, From https://www. linkedin.com/pulse/Android-application-framework-mayur-sojitra (2016)
3. Lessard, J., Kessler, G.C.: Android Forensics: Simplifying Cell Phone Examinations. Retrieved 6 Oct 2018, from https://www.garykessler.net/library/SSDDFJ_V4_1_Lessard_Kessler.pdf (2010)
4. Apple: Mac Technology Overview. Retrieved 9 Oct 2018, from https://developer.apple. com/library/archive/documentation/MacOSX/Conceptual/OSX_Technology_Overview/ CocoaApplicationLayer/CocoaApplicationLayer.html#//apple_ref/doc/uid/TP40001067-CH274-SW1 (2015)
5. Jacobs, B.: Exploring the iOS SDK. Retrieved from https://code.tutsplus.com/tutorials/ exploring-the-ios-sdk–mobile-13959 (2012)
6. Jacobs, B.: Retrieved from https://cocoacasts.com/what-is-the-difference-between-core-data-and-sqlite/ (2016)
7. Proffitt, T.: Forensic Analysis on iOS Devices. Retrieved from https://www.sans.org/reading-room/whitepapers/forensics/forensic-analysis-ios-devices-34092 (2012)
8. Hazra, S., Mateti, P.: Challenges in Android Forensics. Retrieved 20 Nov 2018, from https:// www.researchgate.net/publication/320952681_Challenges_in_Android_Forensics (2017)
9. Mylonas, A., Meletiadis, V., Tsoumas, B., Mitrou, L., Gritzalis, D.: Smartphone Forensics: A Proactive Investigation Scheme for Evidence Acquisition. Retrieved 22 Nov 2018, from https:// link.springer.com/content/pdf/10.1007/978-3-642-30436-1_21.pdf (2012)
10. Lucideus Research: Android Forensic Acquisition Techniques|Lucideus Forensics. Retrieved 23 Nov 2018, from https://lucideustech.blogspot.com/2018/01/android-forensic-acquisition-techniques.html (2018)
11. Murphy, C.: A New Solution for Previously Encrypted iOS Backups. Retrieved from https:// www.gillware.com/forensics/blog/digital-forensics-case-study/new-solution-encrypted-backups/ (2018)
12. Katalov, V.: The art of iOS and iCloud forensics. Retrieved from https://blog.elcomsoft.com/ 2017/11/the-art-of-ios-and-icloud-forensics/ (2017)
13. Satish, B.: (n.d.) Forensic Analysis of iPhone Backups. Retrieved from https://www.exploit-db.com/docs/english/19767-forensic-analysis-of-ios5-iphone-backups.pdf
14. Infante, R.: Reverse Engineering the iOS Backup. Retrieved from https://www.richinfante. com/2017/3/16/reverse-engineering-the-ios-backup (2017)

15. Android Doctor: Samsung Galaxy S7/S7 Edge Root Android 8.0 100% Working. Retrieved 25 Nov 2018, from https://www.youtube.com/watch?v=JwLja68KmKM (2018)
16. IOS Architecture.: Retrieved from https://aruniphoneapplication.blogspot.com/2017/01/ios-architecture.html (2017)
17. Intellipaat: (n.d.). IOS Architecture—iOS Tutorial. Retrieved 9 Oct 2018, from https://intellipaat.com/tutorial/tutorial-ios-tutorial/ios-architecture/
18. McElhearn, K.: Access your iPhone App's data and files. Retrieved from https://imazing.com/guides/how-to-access-your-iphone-apps-files-and-data (2017)

SSD Forensics: Evidence Generation and Analysis

Micah Gibson, Nicole Medina and Zachary Nail

Abstract In this chapter, we forensically examine three different solid state drives (SSDs) from three different manufacturers using three commercial digital forensic tools and one open source forensic tools, and report on the findings of the analysis.

1 Introduction

In today's technology-driven world, computer forensic investigation tools and techniques used to facilitate the discovery and collection of evidence must keep pace with advancements in computer technology. With each new mass-market adopted computer technology, the need to evolve and advance these forensic tools and techniques is clear. One such need for advancement is driven by the rapid market-wide transition from the magnetic hard disk drive (HDD) to solid-state storage (SSD). SSDs are quickly supplanting HDDs as secondary storage in consumer and business computing and entertainment products, including desktops computers, laptops, smartphones, and gaming devices.

While forensic techniques are similar between these devices, SSDs take some extra considerations due to the way SSDs mimic HDD data storage standards. Issues primarily arise in solid-state forensics when the SSD's controller obfuscates drive management from the operating system, especially the SSD's need to "garbage collect" in order to maintain drive performance. This paper will discuss physical and functional differences between HDDs and SSDs, explore forensic considerations that are unique to SSDs, and examine some of the research that has already been performed in these areas before ultimately presenting a brief introduction to the proposed research topic.

M. Gibson (✉) · N. Medina · Z. Nail
Department of Information System and Cyber Security,
University of Texas at San Antonio, San Antonio, TX, USA
e-mail: micahdg@xcomufo.com

© Springer Nature Switzerland AG 2020
X. Zhang and K.-K. R. Choo (eds.), *Digital Forensic Education*,
Studies in Big Data 61, https://doi.org/10.1007/978-3-030-23547-5_11

2   HDDs Versus SSD

Hard disk drives store binary data in the tracks of spinning magnetic disks. HDDs use two types of addressing to organize the data on the drive: cluster, head, sector (CHS) and logical block addressing (LBA). Computer operating systems are able to use these addressing methods to locate data directly on a HDD. The space on a HDD is divided into 512 byte blocks [1]. The magnetic disks in HDDs are capable of handling millions of rewrites over the life of the drive and data can typically be stored safely on a HDD for years without worrying about degradation [1].

Solid state disks use NAND flash memory to persistently store data [2]. SSDs have a controller made up of specialized proprietary hardware and software [3]. The controller has a flash translation layer which translates CHS and LBA addressing to specific pages and blocks in the NAND flash array. Because of this translation, the computer operating system does not know the true location of data in a SSD [3]. Memory is divided into 4–8 k pages and pages are typically organized into blocks 128–256 pages [1]. A block is the minimum sized object that can be deleted. Memory must be erased (all transistors set to 1) before being rewritten [2]. The NAND flash inside of SSDs has a limited number of rewrite operations before the memory becomes incapable of retaining state. Heavy use will lead to bad blocks [4]. To help address the limited life of flash memory, SSD controllers manage what is known as wear-leveling: the active management of write operations so that writes are spread across all memory blocks in an effort to reduce the load on individual blocks [5]. To account for bad blocks, wear-leveling, and the requirement that an entire block must be deleted if any pages in it are to be erased, SSDs have extra blocks not accounted for in the capacity description of the drive, known as over-provisioning. The SSD's controller uses these blocks as if they were the original bad or erased block and simply re-assigns addresses to the over-provisioned blocks when they are used. Likewise, the address of any block can be re-assigned arbitrarily by the controller without the operating system having any idea [6].

3   Forensic Analysis of HDD Versus SSD

Forensic analysis of hard disk drives is fairly advanced today. Because HDDs have been used in computers for decades [3], there are many mature forensic analysis tools capable of lifting forensically-sound images from the disk. Ways that people try to hide or disguise data on HDDs are commonly known among investigators because there are only so many ways a HDD can be utilized to store data [3]. With a HDD, the computer operating system has a good understanding of where data is actually stored on the drive because the CHS and LBA addresses stored in a drive's file allocation table point directly to the physical location of the data on the drive. The file allocation table also knows exactly which blocks are supposed to be empty or no longer used. When data is deleted by the operating system, that data's entry in the file allocation table is deleted but the data is left intact on the drive until the space is needed for future write operations. All of these attributes of HDDs make it easy for forensic tools such as FTK, EnCase, and Autopsy to map existing files and locate anomalous artifacts during investigations [3].

Solid state drive forensic analysis is not as advanced as HDD analysis because of key technological differences between these types of media and differences between SSD implementations by different manufacturers at different times as the technology evolves. There are two main problems and handful of smaller ones that forensic investigators must take into consideration during analysis.

First, SSDs use a proprietary hardware and software controller to mimic the storage activities of a HDD. Computer operating systems, for the most part, issue generic commands to HDDs and SSDs alike during file management. Is the case of SSDs, however, the controller obfuscates what is actually happening to the data stored in the NAND flash array. This mimicry and obfuscation occur in what is also known as the flash translation layer (FTL). The controller's FTL handles a wide of array of extra tasks which must be completed in the flash memory in order for the SSD to have a useful lifespan as a storage medium [7].

The controller's FTL leads to the second main problem for forensic analysis of SSDs: garbage collection or TRIM. Because of the complication of needing to erase an entire memory block in order to erase a single page of data, SSDs would quickly run out of available blocks and then spend an intensive amount of time reorganizing data during future write operations. The TRIM command allows the controller to organize data more efficiently on the SSD at the time data is deleted so that future writes can happen quickly and unabated. Forensically, properly implemented TRIM is a disaster. The TRIM command instructs the SSD controller to move good data out of a block designated for deletion and then promptly wipes the entire block, setting all transistors the same value. This eliminates all chances of recovering deleted data [7]. However, because the SSD market and SSD controller firmware are still in a state of frequent revision by manufacturers and not all operating systems and hardware configurations properly implement TRIM drivers, there are still instances in which deleted artifacts are left behind to be found. TRIM may not work properly if the motherboard, SSD controller, operating system, or the RAID array do not support it or implement it properly. The controller also presents data to the operating system using the Deterministic Read After Trim (DRAT) or Deterministic Zeros After Trim (DZAT) protocols. These protocols cause the controller to return zeros to the operating system for data blocks that have been designated for erasure using TRIM, even when the blocks have not yet been deleted [3].

The SSD's controller also manages wear-leveling, an effort to spread out write operations across all data blocks in the NAND flash array. Data isn't always where the operating system (and by extension, the forensic investigator) thinks it is because the FTL in the SSD controller will move data around and write simultaneously to multiple flash planes in support of wear leveling and the operating system will be none the wiser. This makes accessing the physical blocks almost impossible without a manufacturer's proprietary ATA commands, of which there are many different implementations across the industry [7].

Bad blocks are managed by the SSD's controller. Blocks are marked bad when the SSD has trouble writing to them. When this happens, the controller marks the block as bad, reads the data in it, and moves the data to another block. Bad block marking leaves data vulnerable to forensic analysis because bad blocks cannot be overwritten but can still be read.

Even though an SSD's controller manages the utilization of over-provisioning blocks, format commands and data erasure tools will not always overwrite these over-provisioning blocks either because the controller has a bug preventing it from considering these blocks, or the particular format command or erasure tool doesn't instruct the controller to take these blocks into account. This can leave the over-provisioning blocks accessible to forensic investigators [4].

In another example of SSD controller activity hidden from the computer operating system, some SSDs implement data compression. Without knowledge of the controller's proprietary FTL commands, an investigator trying to access the data directly in the flash memory would not get useful data [5].

Collectively, the data and block management occurring in the SSD controller to mimic HDDs and to extend the life of the SSD's flash memory conspire to make effective forensic analysis of SSDs extremely difficult and sometimes even pointless.

4 Examination of Previous Research

Surveying the existing research into solid-state disk forensics, we find a handful of studies which attempt to examine and compare an array of SSDs in a variety of scenarios.

In Christopher King et al.'s "Empirical analysis of solid state disk data retention when used with contemporary operating systems" [7], the authors perform data recovery and manual TRIM tests on sixteen different models and brands of SSDs on Windows XP, Windows 7, and Ubuntu Linux, with varying densities and usages of file systems. Their findings include that data isn't always where you think it is because the Flash Translation Layer in the SSD controller will move data around and write simultaneously to multiple flash planes in support of wear leveling. Experimentation showed that drives with TRIM enabled and working properly typically wiped all deleted data and left nothing to be recovered by forensic tools. Tools used include Sleuth Kit and python scripts.

J.W. Fulton's "Solid state disk forensics: Is there a path forward?" [6], the author notes that wear leveling, over-provisioning, TRIM garbage collection, all of which can happen simply by plugging in the drive's power. Therefore, forensic soundness is difficult to guarantee from an SSD image because of the constant activity of the SSD controller. The author also notes that FTK and EnCase pull an image of what appears to be a spinning disk because the SSD controller presents the data that way. These tools cannot see through the flash translation layer. The author advocates for a write-protect physical switch on SSDs to allow for more forensically sound image capturing.

In "Black-box forensic and antiforensic characteristics of solid-state drives" by Bonetti et al. [5], TRIM garbage collection simulations are performed by formatting 3 drives and using large files. The authors allowed the drives to sit for multiple hours before examining whether TRIM ran. They also kept drives busy for multiple hours to demonstrate TRIM behavior. The authors found that with TRIM disabled, unavailable, or otherwise broken, data recovery was close to 100%. With TRIM enabled and working properly, zero data was recoverable within seconds of file deletion. The authors also tested and demonstrated the functionality of wear-leveling, compression, erase patterns, and file recoverability.

Belkasoft's lab published a report by Y. Gabanov et al. in which generalized findings from hundreds of different SSDs, acquired while developing and analyzing the behavior of their own data acquisition tool, were summarized. Their findings indicate that SSD forensics can be a "wild, wild west" of computer forensics because of so many different and potentially buggy implementations of garbage collection. Bountiful evidence can be had if TRIM is not working due to driver, firmware, or settings issues or due to some RAID use. The presence of Deterministic Read After Trim (DRAT) or Deterministic Zeros After Trim (DZAT), both of which return zeros instead of the original data after a TRIM command is executed on a data block, also inhibits the efforts of an investigator.

5   Hands-On Project Methodology

For our hands-on research project, we cloned a Window 7 image onto three different SSDs and used three different disk forensic tools—Forensic Toolkit (FTK), EnCase, and Autopsy—to analyze the SSDs. These analyses and our experiences using these tools will be compared to each other and also against an analysis of the original virtual disk image. In the end, we will see that these tools vary widely not only in experience necessary to operate them and the quantity and quality of their results, but also in hardware requirements for a productive forensic analysis.

6   Overview of Source Disk Image

For the purposes of this research project, we procured a Microsoft Windows 7 Ultimate (with Service Pack 1) bootable virtual image to run the forensic tools against. This image was installed in 2013 into a VMware Workstation container on an SSD, but as an HDD, and used almost daily for approximately 4 years. It was primarily used to play a video game in a web browser and to facilitate Skype instant messaging conversations. The virtual image is approximately 20 GB in size.

7 Overview of the SSDs

We used three distinctly different SSDs from three different manufacturers. Our test SSDs and the USB3-to-SATA cable we used are as follows:

Intel SSD 330 Series 180 GB Model#SSDSC2CT180A3 Source: Used for several years and fully formatted prior to this use	[8]
Patriot 120 GB Model #PBU120GS25SSDR Source: Purchased new for this project	[9]
Crucial BX500 120 GB Model #CT120BX500SSD1 Source: Purchased new for this project	[10]
Sabrent 2.5" USB 3.0 to SSD SATA hard drive adapter Model #EC-SSHD	[11]

Our virtual image described in the previous section was cloned onto each SSD using the USB3-to-SATA data and power cable and CloneZilla v2.5.6-22 following instructions found at HowtoForge [12]. The cloned image was placed on each drive in a 30 GB partition, leaving all remaining space unpartitioned and unallocated.

8 Overview of Forensic Tools

The three forensic tools used for this project were Forensic Toolkit (FTK), EnCase, and Autopsy. We first performed a cursory investigation of their main features and compared them in the table below. This is not yet a comparison of the ease of use, usefulness, or thoroughness of any of these tools' features.

	FTK ACCESSDATA [13]	EnCase EnCase Forensic [14]	Autopsy AUTOPSY [15]
Imaging	✓ to a lot of formats	✓	✓ to a vhd file as it scans the drive
Automatic file system detection	✓	✓	✓
Case management	✓	✓	✓
Data carving	✓ Various options based on file signatures	✓	✓
Label, Bookmark	✓	✓	✓
Grouping/classifying investigation achievements	✓	✓	✓

(continued)

(continued)

	FTK ACCESSDATA [13]	EnCase EnCase® Forensic [14]	Autopsy AUTOPSY [15]
Built-in viewers	✓ Hex, Text, Filtered, Natural	✓ Hex, Text, Filtered, Natural	✓ Hex, Text, Filtered, Natural
Filtering and searching	✓ Live search, index search	✓	✓ Keyword search with regex
File decryption	✓	✓	✗
Visualization	✓	✓	✓
Timeline	✓	✓	✓
Report	✓	✓	✓

We also researched pricing options for these tools:

	FTK AD ACCESSDATA	EnCase EnCase® Forensic	Autopsy AUTOPSY DIGITAL FORENSICS
Per User or Workstation	$3,995 (perpetual) [16] $2,227 (1 year of license + support) [16]	$2,995.00 [17]	Free
Support (1 year)	$1,119 [16]	$688.85 [17]	Unknown
Training	$2,495 [18]	$5,085–$6,495 [19]	$499 [20]

9 Experiences During Evidence Drive

9.1 Forensic Toolkit

The Forensic Toolkit (FTK) v6.0.3.5, sold and maintained by AccessData, is available with various licensing schemes [16] and is considered by many to be the best hard drive and SSD forensics tool in the industry [21]. Using FTK, we analyzed the Crucial SSD and the original VMware virtual image separately. The original image was analyzed as a baseline to which we could compare all other evidence acquisition and analysis.

FTK was found to be powerful and thorough, able to easily and quickly organize evidence by file type, extension, location on the drive, allocation status, and more. It was also readily able to carve out deleted and partially overwritten data as well as search for social security and credit card numbers. The counts of files and other search results are included in the evidence drive tables later in this paper.

We had some challenges with FTK. Processing the Crucial SSD as new evidence was very time consuming, taking around three hours to complete. FTK doesn't automatically find and categorize internet browsing history, cookies, or browsing cache. It likewise could not find an internet keyword search history. FTK also was not able to read database file contents for purposes of string searching. It also categorized $MFT entries and attributes as mismatched extensions which led to a skewing of the carved data numbers. The tool for automatic file decryption requires a user-provided password list to run against the encrypted files. Additionally, because FTK is an expensive application, we were forced to use the copy available in the computer lab and deal with the limitations that entailed. These limitations included the availability of only one type of PC hardware configuration and unreliable local evidence storage. We had to utilize the network share to store the approximately 70 GB of data because we could not always guarantee we would be able to use the same workstation during each working session. The CPU in these workstations was not very powerful, which led to maxed out CPU cores during most of the time FTK was in use, causing the workstations to freeze up frequently for 10+ min at a time. These computer lab workstations were also limited by their old USB 2.0 interfaces. Analyzing the SSD for the first time and during subsequent keyword searches proved to be very slow and take hours for each search. Since we could not run this software outside the computer lab, we don't know whether the freezing issue was the software itself (perhaps not as optimized as it could be), the bottleneck of the USB 2.0 interface, the slow CPU, or our reliance on network storage for the artifact repository.

When used to collect evidence from the baseline virtual image, FTK's file carving proved to be much more fruitful than when used to collect evidence from the SSD. This is presumably because the SSDs' controller had probably used TRIM and other processes to clean up and remove more of the evidence before or even while FTK examined the drive.

The pricing for FTK was not easily determined. Precise costs require talking directly with an AccessData representative and working out a deal with them. We were able to find some price estimates from a third party that broke it down by only single perpetual licenses or single yearly licenses [16].

9.2   EnCase

EnCase® Forensic

EnCase v7.12.01.18, distributed and maintained by Guidance Software/opentext, is another industry-leading hard disk and SSD forensics analysis tool. This tool was also only available to us in the computer lab.

EnCase was by far the most difficult tool to use. It suffered, more than FTK, from the USB 2.0 interface bottleneck and other computer lab workstation limitations. Whereas FTK was able to consume all CPU cores, EnCase was single-threaded and limited to only one core, leading to many time consuming freezes and delays while attempting to process the Patriot SSD. Sadly, we were not able to make a local image of the data using EnCase, so each time we wanted to search through the data for more evidence, the application would freeze for long periods of time. Additionally, due to a lack of hard disk space on some of the workstations and our inability to exclusively use the same workstation during each session, the forensic result data was stored on a network drive that also caused latency issues. We also struggled with getting EnCase to remember where the case file data was stored and that it belonged to the Patriot SSD. This problem was further exacerbated by EnCase hardcoding paths to its data which presented problems when the network path's or Patriot SSD's drive letters changed, which they did with each different lab workstation.

It was, however, possible to perform a filter search and clearly see every item associated with that category. The given list marked any additional information on a thumbnail image next to each item, indicating the results of a file signature analysis and whether the file was deleted and/or fragmented. While this information appears useful on a per file basis, it was difficult to aggregate the results for further analysis. EnCase's case management system was easy to use and automatically tracked and saved searches.

The graphical user interface in EnCase was very difficult and confusing to use. There are hundreds of boxes to check, buttons to push, and dropdowns to select from, and almost none of them explain what they control. Google searches for explanations and guides for keyword searches often proved fruitless because the instructions were for a version of EnCase that was too different from the version we had in the computer lab, which led us frequently to dead ends where we could not figure out what to click next to finish what we wanted to do. Trying to intuitively click around did not accomplish anything useful. There is very expensive official training available for EnCase which includes several days of hands-on instruction. We wonder if the interface is difficult to use in order to spur more customer spending on training and instruction.

The deleted files that EnCase was able to find were mixed in with the undeleted files, making it not very straight forward to tell whether someone had tried to delete a file.

Similar to FTK, pricing information for EnCase was not readily available from the developer, Guidance Software. Instead, third party retailers were consulted to get pricing estimates for the perpetual license and first year of support [20].

9.3 Autopsy

The Sleuth Kit and its associated graphical user interface, Autopsy v4.9.1, currently maintained by Basis Technology, is available as open source. This open source aspect of Autopsy proved to be invaluable in providing extra time to learn to use the tool and computing power to analyze the Intel SSD.

Unlike with FTK and EnCase, we were able to use our own computers, with their USB3 interfaces, speedy SSDs, and more RAM, to make quick, short work of the Intel SSD analysis. The processing of this SSD took about an hour and Autopsy generated a virtual disk image as it went so that it would only need to read from the SSD once. Autopsy also kept the user interface with all its results constantly available and updated as it made progress during the analysis. This was very useful for providing immediate results to validate that everything was set up properly. Autopsy populated lots of useful categories of data without user input, including carved and deleted files, web history, phone numbers, and potential zip bombs. Its case file organization was also straightforward and easy to use.

Overall we found no particular shortcomings with the Autopsy software. Basis Technology also offers to assist in writing any add-on modules a customer may need for their particular forensic investigatory use. However, an organization using Autopsy must have its own in-house Autopsy experts because there is no paid usage support available.

9.4 Evidence Drive Results

For our research project's evidence drive, after each tool processed the SSD images and gave us a general report of their findings, we found many instances in which each tool was looking for specific categories and types of files. We combined these results to create many of the below tables of evidence. Some of the tables, especially the keyword and specific file searches, required the creation of search term lists or manual searches by the investigator.

Files of interest***	Carved files #	Orphaned files #	Encrypted files #
FTK—baseline	111,774	35,769	40
FTK	56,497	30,013	13
EnCase	1,043*	15,774	3
Autopsy	3,198	14,998	13

*This represents the carving of only documents, emails, multimedia, and pictures. When more thorough searches were attempted, EnCase would stop responding and not recover before the computer lab closed
***Files of interest are files which warrant closer examination if an investigator is looking for deleted and encrypted data

File types	Videos #	Images #	Audio #	Zip/Archive files
FTK—baseline	164	10,554	687	1,061
FTK	164	10,519	664	1,061
EnCase	136	10,329	677	736
Autopsy	164	7,976	656	105

File types	HTML/HTM #	PDF #	Doc #	Text/Rich text #
FTK—baseline	1,069	3	1	1,375
FTK	1,069	3	1	1,372
EnCase	1,069	3	3	1,373
Autopsy	652	0	2	2,492

Recycle Bin findings	# SIDs	# files	Deleted files #
FTK—baseline	1, user #1000	43	49,505
FTK	1, user #1000	43	47,946
EnCase	0	4	35,553
Autopsy	1, user #1000	15	62,166

Web history findings	History entries	Web downloads	Cookies	Search history
FTK—baseline	46,053	9	822	n/a
FTK	42,631	9	801	n/a
EnCase	1,410	2	1,069	43 terms
Autopsy	52,088	14	792	36 terms

Other	Extension mismatches	Email addresses	Zip bombs	User accounts found
FTK—baseline	61,415	733	n/a	3—kyle, guest, and Administrator
FTK	43,615	664	n/a	3—kyle, guest, and Administrator
EnCase	0	1,077	813	2—kyle and Guest
Autopsy	17	17,862	813	1—Username: kyle

Other	# of events over lifetime of drive	Credit card numbers	Phone numbers	SSNs
FTK—baseline	Not available	34	6,862	5
FTK	Not available	31	6,694	4
EnCase	Not available	**	**	**
Autopsy	439,697	0	516	0

**Indicates software or hardware limitations that prevented us from completing the searches within a timely manner or within 3 h each before being kicked from the computer lab

Keyword search	"Dark web"	"tor"	"Hiding body"	"Hidden"	"bmap"	"metasploit"	"Gmail"
FTK–baseline	**	**	**	**	**	**	**
FTK	**	**	**	**	**	**	**
EnCase	**	**	**	**	**	**	**
Autopsy	0	581	0	3,664	69	1	327

Specific app search	Is teamviewer installed?	Is opera installed?	Is civilization installed?
FTK—baseline	Yes	Yes	Yes
FTK	Yes	Yes	Yes
EnCase	**	**	**
Autopsy	Yes	Yes	Yes

10 Analysis and Summation of Results

The most interesting results of the evidence drive was the difference between the data gathered from the baseline image and the cloned SSDs. In the vast majority of the evidence drive reporting categories, the baseline image yielded greater numbers of artifacts than what could be pulled from the SSDs. We believe these differences are due to the SSDs' controllers actively trying to more efficiently arrange the cloned data by utilizing TRIM, wear leveling, and other proprietary data management techniques. By simply applying power to the SSDs during the evidence retrieval process, the controllers were allowed to clean up data that had been cloned onto the drive but previously marked for deletion while the baseline image was last running, resulting in a dramatic reduction in located artifacts in many categories.

Within the limited scope of our project, Autopsy was consistently able to deliver the types of evidence we were searching for. It may not have included all the data FTK was able to find, but it at least returned results in all categories. Of course, this is a double-edged sword because it could lead to a false assumption that Autopsy was able to find all the artifacts that can be found and that no more thorough searching should be conducted.

Due to the limitations we encountered while trying to use the computer lab's copies of FTK and EnCase, we feel that our report may not truly encompass the breadth of capabilities these forensic tools have to offer. Vassil Roussev et al. discussed in 2004 the growing need for distributed processing of forensic data in order to overcome the bottlenecks and limitations of a single workstation [22]. Even though their research was performed fourteen years ago, the amount of data that must be processed and categorized during forensic investigations today has only exponentially increased since then. If we were able to run these applications on our own superior hardware, or on some of the purpose-built forensic hardware available on the market today [23], or perhaps even utilize special-purpose cloud computing using AccessData's AD Lab and Amazon Web Services [24], we believe we would have gotten faster and more thorough results from FTK and EnCase.

References

1. Solid State/Hard drive differences: D F I News. https://search-proquest-com.libweb.lib.utsa. edu/docview/1626732768 (2014)
2. Micron: An introduction to NAND flash and how to design it in to your next product. NAND Flash **101**, 27. https://ece.umd.edu/~blj/CS-590.26/micron-tn2919.pdf (2006)
3. Gubanov, Y., Afonin, O.: Recovering evidence from SSD drives: understanding TRIM, garbage collection, and exclusions. D F I News. https://search-proquest-com.libweb.lib.utsa. edu/docview/1613504182 (2014)
4. Crump, G.: Flash storage has special security needs. Informationweek. https://search-proquest-com.libweb.lib.utsa.edu/docview/1437266763?accountid=7122 (2013)
5. Bonetti, G., Viglione, M., Frossi, A., Maggi, F., Zanero, S.: Black-box forensic and antiforensic characteristics of solid-state drives. J. Comput. Virol. Hacking Tech. **10**(4), 255–271 (n.d.). https://doi.org/10.1007/s11416-014-0221-z
6. Fulton, J.W.: Solid state disk forensics: Is there a path forward? (Order No. 1555597). Criminal Justice Database; ProQuest Dissertations and Theses Global (1534359347). https://search-proquest-com.libweb.lib.utsa.edu/docview/1534359347?accountid=7122 (2014)
7. King, C., Vidas, T.: Empirical analysis of solid state disk data retention when used with contemporary operating systems. Digit. Investig. **8**(2011), S111–S117. http://www.sciencedirect.com/science/article/pii/S1742287611000375 (2011)
8. Intel SSD Image: Intel SSD 330 series 180 GB. https://images-na.ssl-images-amazon.com/images/I/71VVSWNNxoL._SX425_.jpg. Accessed 14 Nov 2018
9. Patriot SSD Image: Patriot 120 GB. https://c1.neweggimages.com/NeweggImage/ProductImage/20-225-081-Z01.jpg. Accessed 14 Nov 2018
10. Crucial SSD Image: Crucial BX500 120 GB. https://img.bfmtv.com/c/630/420/b17/f8f244eccb926c4f398ed91c680fc.jpg. Accessed 14 Nov 2018
11. Sabrent USB-to-SATA Adaptor Image: Sabrent EC-SSHD 2.5" USB 3.0 to SSD SATA Hard Drive Adapter. https://www.sabrent.com/uploads/EC-SSHD-Main.jpg. Accessed 14 Nov 2018
12. HowtoForge: Converting a VMware Image to a physical machine. https://www.howtoforge.com/converting-a-vmware-image-to-a-physical-machine. Accessed 20 Oct 2018
13. AccessData Logo Image: https://accessdata.com/assets/images/logo.png. Accessed 14 Nov 2018
14. EnCase Logo Image: https://www.guidancesoftware.com/docs/default-source/brand/ef-lockup-2colorba91e5f7beb06108b3ffff0100c79052.png?sfvrsn=2. Accessed 14 Nov 2018
15. Autopsy Logo Image: https://www.autopsy.com/wp-content/uploads/sites/8/2016/02/autopsy-logo-head.svg. Accessed 14 Nov 2018
16. SC Media SC Magazine: Product information AccessData Forensic Toolkit (FTK), 3 October 2016. https://www.scmagazine.com/review/accessdata-forensic-toolkit-ftk/. Accessed 14 Nov 2018
17. Digital Intelligence: EnCase forensic V8 + 1 year subscription. https://digitalintelligence.com/store/products/s5300. Accessed 14 Nov 2018
18. Tittel, Ed., Lindros, K.: Best Digital Forensics Certifications. https://www.businessnewsdaily.com/10755-best-digital-forensics-certifications.html. Accessed 14 Nov 2018
19. opentext: Training Packages. https://www.opentext.com/products-and-solutions/services/training-and-learning-services/course-catalogue/encase-training/training-packages. Accessed 14 Nov 2018
20. Autopsy: Training in autopsy. https://www.autopsy.com/training/. Accessed 14 Nov 2018
21. Choo, R.: AccessData Forensic Toolkit (FTK) v6.4 lecture slides. https://utsa.blackboard.com/bbcswebdav/pid-3326004-dt-content-rid-51345344_1/xid-51345344_1. Accessed 20 Oct 2018
22. Roussev, V., Richard III, G.G.: Breaking the performance wall: the case for distributed digital forensics. In: Proceedings of the 2004 Digital Forensics Research Workshop, vol. 94. http://www.academia.edu/download/30684064/Golden-Perfromance.pdf (2004). Accessed 14 Nov 2018

23. Forensic Computers Incorporated: Forensic tower IV X10 dual xeon. https://www.forensiccomputers.com/forensic-tower-iv-x10.html. Accessed 14 Nov 2018

24. AccessData: AD LAB: large-scale investigation and processing. https://accessdata.com/products-services/ad-lab. Accessed 14 Nov 2018

25. Black, M.E.: The impact of solid-state drives on digital forensic investigations: a case study (Order No. 3737151). ProQuest Dissertations and Theses Global (1747435217) (2015). https://login.libweb.lib.utsa.edu/login?url=https://search-proquest-com.libweb.lib.utsa.edu/docview/1747435217?accountid=7122

26. Cornwell, M.: Anatomy of a solid-state drive. Commun. ACM, 55(12), 59–63 (n.d.). https://doi.org/10.1145/2380656.2380672

27. Gubanov, Y.: Retrieving digital evidence methods, techniques, and issues, 14 June 2016. https://www.forensicmag.com/article/2012/05/retrieving-digital-evidence-methods-techniques-and-issues. Accessed 14 Oct 2018

28. Remo Software: How to choose the most effective SSD data recovery technique. Remo Hard Drive Recovery Software, 6. https://www.remosoftware.com/ssd-data-recovery-techniques (2018)

29. Micheloni, R.: Solid-State Drive (SSD): a nonvolatile storage system. Proc. IEEE **105**(4), 583–588 (n.d.). https://doi.org/10.1109/JPROC.2017.2678018

Web Browser Forensics in Google Chrome, Mozilla Firefox, and the Tor Browser Bundle

Rebecca Nelson, Atul Shukla and Cory Smith

Abstract Browsers are widely used on personal computers, laptops and mobile devices. In this chapter, we seek to determine and compare which forensic artifacts can be recovered from Google Chrome, Mozilla Firefox, their respective private modes, and TOR. Our analysis was primarily conducted using FTK in order to replicate the process and abilities of a digital forensics lab with limited resources. After identical data generation across all browsers and modes of browsing in a controlled virtual environment, forensic images were captured then analyzed. This research not only extends the current field of digital forensics for which artifacts can be found in which locations, but also confirms various claims in regards to the privacy of private browsing modes. As expected, all data was recovered from regular browsing modes, very minimal data from private browsing, and almost no artifacts from TOR.

Keywords Web browser forensics · Google Chrome · Incognito · Mozilla Firefox · Private browsing · TOR · The onion router network

1 Introduction

As technology becomes increasingly integrated into our daily lives and we become more dependent on the Internet, many people are becoming more cognizant of their digital footprint and the associated demand for privacy. Regardless of how you connect to the Internet—whether it be from desktop, laptop, or mobile device—a browser is usually required. Web browsers allow users to access news websites, pay bills, watch videos, and send emails. For some, this network traffic may be sensitive in nature. For others (such as hackers or malicious users) it can generate incriminating evidence. The release of thousands of top secret NSA documents in late 2013 caused a massive interest in public information security and the right to privacy [10]. Enter the rise of private browsing capabilities.

R. Nelson · A. Shukla (✉) · C. Smith
Department of Information System and Cyber Security,
University of Texas at San Antonio, San Antonio, TX, USA
e-mail: atul.shukla@my.utsa.edu

© Springer Nature Switzerland AG 2020
X. Zhang and K.-K. R. Choo (eds.), *Digital Forensic Education*,
Studies in Big Data 61, https://doi.org/10.1007/978-3-030-23547-5_12

219

Private browsing allows the user to browse the web without leaving (much) evidence behind about what sites were visited (history), accounts used (autofill, saved passwords), files downloaded, etc. Based on the browser and configurations used, some of this data can changed or eliminated [15]. While some private-browsing features started popping up in the early 2000s, the leaked NSA documents contributed to the explosive growth of privacy-focused browsers, plugins, and addons. These features are quite attractive to regular users, hackers, or criminals for various reasons. However, the usage of private browsers can complicate evidence gathering techniques and processes used by forensic investigators.

Our aim is to extend the study of web browser forensic analysis research to cover current browser versions. Our research seeks to identify which forensic artifacts are left behind when someone browses the web with private browsing mode enabled. We will compare these 'private' artifacts with the artifacts left behind after using default browser settings. Our scope will include Google Chrome, Incognito mode, Mozilla Firefox, and Private Browsing mode. In addition to comparing the regular and private browsing modes of these 2 browsers, we will compare recovered artifacts between private browser modes and the inherently private TOR browser. Our analysis will be conducted using tools commonly used by law enforcement agencies around the nation instead of proprietary or application-specific file carvers. The purpose of this approach is to investigate which artifacts can be recovered after private browsing in support of a civil or criminal investigation; FTK is considered to be one of the primary tools used by law enforcement for this type of digital investigation. In the next section we will discuss previous research in this field, we will then define the data generation process as well as the tools used throughout our research, then we will finish with a detailed discussion of our findings, our overall conclusions, and a list of future research topics.

2   Related Work

Google Chrome is a freeware web browser, and was released in 2008. Chrome is one of the leading web browsers today and has over half of the market share of browser users on both desktop and mobile platforms. Chrome, like other similar browsers that are popular, has a privacy mode called Incognito Mode. Incognito Mode allows the users to browse without the browser storing any of their information to help keep things private in their browsing needs. Chrome's Incognito Mode allows the user to have this privacy after only a few quick clicks.

Shaftqat [23] discusses some of the major differences between Chrome's normal browsing mode and Chrome's Incognito browsing mode. Some of their findings included that in private mode (Incognito), the browser still stored some of the user's information, just in another location. Cookies and bookmarks are stored in the default Chrome location, while history is only stored in RAM. Further, user names and passwords are not stored at all. With some of the user's data still available to discover and find, is Incognito really that private?

Rathod and Digvijaysinh [21] along with Shafqat and Narmeen [23] discuss the tools and methods that we can use to analyze the difference between normal and private browsing modes. There are many ways that you can browse through the files stored on your computer and in your web browser to see the user's data. Some of the forensic tools that were used during these investigations included Chrome-Cookie View, Chrome Password Decryptor, ChromeCache View, Chrome Analysis, and many others.

This is very similar to the research that we are going to be conducting on our Chrome browser. We will discuss how much Incognito Mode and other private browsers hide the user's information. We will be performing tests similar to the ones found in the articles mentioned above on both normal and private browsing modes and compare them to each other, and then we will be able to state our own conclusion.

Mozilla Firefox is one of the top 3 most popular browsers in use today. Mozilla Firefox is an open source browser focused on performance and privacy, while keeping things simple and easy to use [16]. The Firefox browser differentiates itself from other browsers by maintaining a balance between being transparent and secure. Firefox's Private Browsing mode also serves as the foundation for the TOR browser.

Using specialized tools, Oh et al. [19] were able to recover various artifacts and research how integrated web browser forensics can be performed across different browsers. A new tool introduced in their research, Web Browser Forensics Analyzer or WEFA, may not be available to forensic investigators, but the research effectively identifies areas where existing tools may have been lacking. With this information, our research can observe how the tools have advanced over the years and if the capabilities have caught up with the needs of the digital investigators.

Gabet's [7] research is most closely related to our research objective. While Gabet focuses on recovering artifacts from three private browsers compared to private *modes* of three common browsers, other works focus on analyzing disk space [20] or a single browser's artifacts [12, 21]. As mentioned earlier, the tools available seem to be lacking in one aspect or another. Many works choose to focus on tool comparisons by conducting similar browsing steps with different recovery methods, to see which tools recover which or how many artifacts [13, 17, 19]. Another direction that some researchers have taken is to focus on memory dump analysis and associated memory or file carving techniques with various tools to see which browsers leave behind which artifacts [8, 18]. This method, in combination with Gabet's method, is where our research begins—analyzing private browser artifacts across a few popular browser choices. It's important to keep in mind that most artifacts recovered in other works were done so with browser-specific or process-specific tools; we will only utilize AccessData's Forensic Toolkit.

Out of all previous and related works, Tsalis et al. [25], take one of the more unique approaches not covered by other research efforts before. The group focused on which artifacts can be recovered from an attacker's perspective with limited technical forensic knowledge and no special tools. This practical approach focuses only on the file system of a machine after the browser was closed, but before it was powered off. If an attacker were to gain remote access to a workstation, what would they be able to find and uncover without tipping off the end user or using complex carving tools.

The Tor Project has become a well-known browser for those seeking anonymity on the web [24]. There are many driving forces for using the Tor Browser Bundle (TBB) for people all over the world; the Tor Project, which is responsible for promotion and development of the Tor network, suggests the following reasons for using the TBB: ensure privacy from online marketers, secure communication across unsecured networks, protecting children online, sensitive research, as well as avoiding surveillance and censorship. Tor is also popular in the law enforcement, activists, journalists, and business executive circles.

The Tor Project provides detailed instructions for using the Tor browser as well as discusses the ease of using the TBB, and the benefits it can provide its users [24]. The TBB can easily be installed in under 5 min, and provides anonymous browsing instantaneously. The TBB deletes browser cookies and the list of visited sites when the browser is closed, providing anti-forensics techniques to Tor users; because of this, the TBB is a popular browser for those conducting malicious activities.

Several researchers have conducted Tor Browser forensics in an attempt to determine if traditional browser artifacts could be recovered from a machine in which the Tor Browser was used [3, 6, 12, 22, 26]. However, many of these articles focused on determining if any remnants of the Tor browser could be recovered from the machine, not necessarily artifacts created from a browsing session. All articles recovered artifacts from the prefetch file, windows page file, thumbnail cache, and registry that proved that the Tor Browser was in use. Boggs et al., were able to recover visited pages by dumping the contents of the RAM.

Although several attempts have been made to perform a forensic investigation against the Tor browser, they have been on older TBB versions or older versions of the Windows OS. Our research project will focus on Windows 7 and the latest version of the TBB (v7.0.5). The results from this investigation will then be compared to those of similar fashion to the private modes on both Chrome and Firefox.

3 Environment Setup

Our research focuses on retrieving artifacts from popular browsers such as Google Chrome, Mozilla Firefox, and their respective private-browsing modes, and then comparing recovered artifacts with those which can be recovered from Tor browser. While other related works conducted similar experiments, there have not been any recent studies comparing artifacts from updated browser applications with Tor browser, especially in recent years where information security and data privacy has become such a common cause for concern. Our analysis will be conducted using tools commonly used by law enforcement agencies around the nation instead of proprietary or application-specific file carvers. By doing so, we can provide a practical example of which files and artifacts can be recovered by most agencies or investigators who have access to a particular device in regards to browser activity.

In order to get reliable and realistic data for browser artifact comparison, each browser had two identical virtual machines. The only difference was that one machine would have default browser settings while the other machine was secured via browser settings and advanced privacy settings available to any user. Tor browser, by nature, is already secure and focused on privacy, so this browser only needed one machine. The virtual machines were used only for browsing research and no other activities. Any data, files, or artifacts stored on the virtual machines would have come from the respective browsers and no other applications. Each team member had their own virtual machines on their own host machine in a controlled lab environment, and no one except for the team member and computer lab manager had access to a particular user's virtual machines.

Our data generation process included known 'normal' and 'abnormal' browsing activity over the course of multiple weeks. Each time a browsing session concluded, the browser was closed, virtual machine was shut down, and then a virtual machine snapshot image was saved. Browsing steps and data preservation steps were identical across browsers and virtual machines. Any minor deviations from the data generation playbook were caused by the inability to access a page or perform an action due to the security features of that browser. Based on the situation, an equivalent action was performed, or no action was performed; any deviation was documented.

Before imaging the virtual machines for forensic analysis, VMware Virtual Disk Manager was used to aggregate and compile all snapshot vmdk files into one vmdk which was then used in FTK Imager. Once the virtual machine snapshots were imaged through FTK Imager, analysis was performed on the image files to determine which browser artifacts can be recovered from browsers at varying levels of security and privacy.

Our hypothesis is that there will be fewer useful and recoverable artifacts from Tor browser when compared to private-mode browsing on other browsers. In addition, private-mode browsers will produce fewer recoverable artifacts when compared to browsers on default settings. Therefore, private-mode browsing, or Tor when possible, will provide a more secure and private browsing experience for the end user. However, this will make digital forensic analysis of these users much more difficult without the use of specialized or application-specific tools.

Table 1 provides detailed information regarding the tools used, their version and the process of generating the transactions to be investigated to allow for easy replication and verification of our research and findings. Several tools were used to ensure a solid framework for this research.

VMWare Workstation served as the type II hypervisor for this research, on which we hosted identical virtual machines with the following configurations:

- Windows 7 x64 ISO file configured with:
  - 12 GB Memory
  - 2 Processors
  - 75 GB Hard Disk.

Table 1 Tools used throughout research

Tools	Versions
VMWare workstation 12 pro	v12.5.7 build-5813279
Forensic toolkit (FTK)	v6.2.0.1026
FTK imager	v4.1.0.12
FTK registry viewer	v2.0.0.7
WinHex	V19.3 SR-4 ×86
SQLite manager Firefox addon	0.8.3.1

Table 2 Browser versions

Browsers	Version
Google Chrome	v61.0.3163.100 (Official build) (64-bit)
Mozilla Firefox	v55.0-2 (32-bit)
Tor browser bundle	v7.0.5

The Forensic Toolkit suite from Access Data served as the tooling for performing the recovery of the forensic artifacts. FTK Imager was used to create the forensic image, a bit for bit copy, of the vmdk file. Once created, the forensic image was mounted into FTK. Once mounted into FTK, we could interrogate the file systems of our virtual machines. FTK Registry Viewer is a specialty tool used to extract the Windows registry.

In order to conduct a thorough analysis of artifacts recoverable from a private browsing session, we wanted to utilize several different browsers in order to compare and contrast our findings. The browsers, along with their versions, can be seen in Table 2.

In order to generate traffic and capture data to analyze as part of our research, mini playbooks were created for the team to follow during each browsing session. These steps mirrored many common activities that regular users and hackers alike may take while browsing the Internet. Based on the scope of browser artifact forensics, our activities spanned the range from browsing news sites, watching videos online, accessing course content through an education portal, sending and receiving email messages, and creating, downloading, and uploading files to cloud storage. In addition, data was also generated using secure and private search engines, and attempting to disguise traffic through a free online proxy service; both activities may be used by potential hackers trying to hide their tracks. Figure 1 shows a high-level overview of our project workflow.

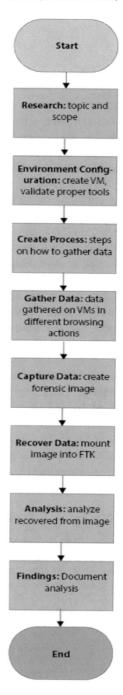

Fig. 1 Research process

4 Findings

4.1 Chrome Analysis

Google Chrome, like many other web browsers, offers a private browsing mode known as Incognito. Many web browsers store data and files on the user's computer and in the browser's history about the sites they have visited, preferences for certain sites and other such browsing traffic data. How Google Chrome works is it "has been built with two separate modules called protection domains: the browser kernel which exchanges messages with the operating system and a rendering engine which runs with limited privileges in a sandbox. The main aim of Chrome is to prevent a malware from being run through the browser and file theft" [18]. Chrome also stores its data in a SQLite database, similar to other web browsers, which you will see examples of below. Chrome's Incognito mode helps remove and block most of this private data, such as cookies and browser history, so that there are less traces being stored within the browsers data.

For this project we gathered data from a few different web browsers but chose Google Chrome as one of our main browsers because it is the most widely used and popular browser in the world. Many articles state that Chrome's Incognito mode is one of the best private browsers. Through our research we will come to see if that is a true statement as we will compare Chrome, Firefox, and Tor's data gatherings. To generate browser traffic and data, we performed identical steps on both a normal Chrome browser and an Incognito browser on separate virtual machines. After taking snapshots of the VMs, compressing them into an image and running them through FTK Imager, we were able to look into the data that was generated through this web browser. Most of the data was found in the following file path: ../root/Users/FTKuser/AppData/Local/Google/Chrome/. There were a number of different artifacts gathered between normal and private browsing (Figs. 2 and 3).

Fig. 2 Normal browsing

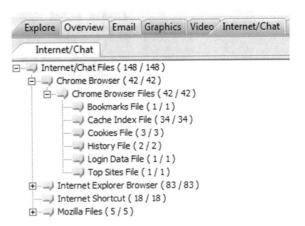

Fig. 3 Private browsing
incognito mode

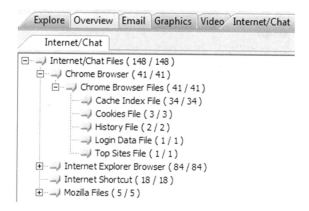

There are many user preference settings in Google Chrome that can make the browser more secure that most people should consider using. A few simple option changes can prevent the browser from storing a decent majority of the user's information and browsing history. In Google Chrome's settings there is a privacy section where the user can disable their browsing traffic from being stored and, "Send a 'Do Not Track' request with your browsing traffic". By default a large majority of websites gather the user's browsing traffic and this option would help by trying to block as much traffic as it could. This however does not always do the trick, because many websites have found loopholes around this and are still able to gather some data. Browsing traffic was one artifact that we investigated on our VM's. Where we used the normal Chrome browser, we did not set any privacy settings, and browsed traffic like normal. We were able to see about 350 different websites that had been visited. Not only were we able to see the site and the URL that was visited, but we were able to see how many times that site was visited, the last time the site was visited and the duration of that visit for all 350 sites. On the other VM with the Incognito Browser, we set the browser's settings to the privacy setting mentioned above. With that setting and with being in incognito, the browser had zero traces of browser history visits. There is a stark difference between the data gathered between the two browsers for the sites visited (Figs. 4 and 5).

With blocking the browsing traffic from the privacy section in settings, there were also other big finds that we noticed while analyzing the browsers data and that was browsing cookies. In the privacy section there is also an option to block cookie data generation such as: "Allow local data to be set", "Keep local data only until you quit your browser", "Block sites from setting any data", and "Block third-party cookies and site data". The last option to block third-party cookies is the best place to start. Third-party cookies are where many advertisers and other companies gather users browsing data. Blocking third-party cookies are one of the most preventative web-tracking methods. While looking at cookies from the data gathered on the Chrome browsers, we saw very similar results as we did with the browser traffic. While in

Chrome History SQLite
Visits

Visit RowId	URL	Redirect URL	Title	Visit Times	Typed Times	Is Hidden	Last Visit	This Visit	Duration
1	https://www.google.com/	[NULL]	Google	9	9	no	9/28/2017 5:36:58 PM -0500	7/19/2017 3:27:31 PM -0500	11 seconds
2	https://www.google.com/search?site=&source=hp&q=firefox&oq=firef&gs_l=psy-ab.1.0.0i131k1l2j0i131i46k1j46i131k1l2.349.1782.0.359 6.5.5.0.0.0.0.109.289.3j 1.4.0....0...1.1.64.psy-ab..1.4.28 8...0.-gbj3Gj4cTg	https://www.google.com/	firefox - Google Search	1	0	no	7/19/2017 3:27:36 PM -0500	7/19/2017 3:27:36 PM -0500	seconds
3	https://www.mozilla.org/en-US/firefox/	https://www.google.com/	A better, faster, private browser for today	Firefox	1	0	no	7/19/2017 3:27:43 PM -0500	7/19/2017 3:27:43 PM -0500	0 seconds
4	https://www.mozilla.org/firefox/new/?scene=2	https://www.mozilla.org/en-US/firefox/	Download Firefox — Free Web Browser — Mozilla	1	0	no	7/19/2017 3:27:47 PM -0500	7/19/2017 3:27:47 PM -0500	0 seconds
5	https://www.mozilla.org/en-US/firefox/new/?scene=2	https://www.mozilla.org/firefox/new/?scene=2	Download Firefox — Free Web Browser — Mozilla	1	0	no	7/19/2017 3:27:47 PM -0500	7/19/2017 3:27:47 PM -0500	23 seconds
6	https://www.google.com/search?q=UTSA&oq=UTSA&aqs=chrome..69i57j0l5.1695j0j8&sourceid=chrome&ie=UTF-8	[NULL]	UTSA - Google Search	1	0	no	8/17/2017 9:29:35 AM -0500	8/17/2017 9:29:35 AM -0500	0 seconds

Fig. 4 Normal browsing visits

Chrome History SQLite
Visits

Visit RowId	URL	Redirect URL	Title	Visit Times	Typed Times	Is Hidden	Last Visit	This Visit	Duration

Searches

Search RowId	URL	Terms

Downloads

Download RowId	Current Path	Target Path	URL	Start Time	State	Bytes Downloaded	Bytes Total	End Time	Opened	Redirect URL

Fig. 5 Incognito visits

Chrome Cookies SQLite

Cookies

Cookie RowId	Host Key	Name	Value	Path	Creation Time	Expiration Time	Last Access Time	Is Secure	Is HTTP Only
13144969667444617	.doubleclick.net	id		/	7/19/2017 3:27:47 PM -0500	7/19/2019 3:27:47 PM -0500	10/2/2017 9:02:15 PM -0500	no	no
13144969667603500	.mozilla.org	optimizelyEndUserId		/	7/19/2017 3:27:47 PM -0500	7/17/2027 3:27:47 PM -0500	7/19/2017 3:27:47 PM -0500	no	no
13144969667610923	.mozilla.org	optimizelySegments		/	7/19/2017 3:27:47 PM -0500	7/17/2027 3:27:47 PM -0500	7/19/2017 3:27:47 PM -0500	no	no
13144969667710111	.mozilla.org	optimizelyBuckets		/	7/19/2017 3:27:47 PM -0500	7/17/2027 3:27:47 PM -0500	7/19/2017 3:27:47 PM -0500	no	no
13144969667809011	.mozilla.org	_ga		/	7/19/2017 3:27:47 PM -0500	7/19/2019 3:27:47 PM -0500	7/19/2017 3:27:47 PM -0500	no	no
13144969667967921	.246059135.log.optimizely.com	end_user_id		/	7/19/2017 3:27:47 PM -0500	7/17/2027 3:27:47 PM -0500	7/19/2017 3:27:47 PM -0500	no	no
13147453780985421	.bat.bing.com	MR		/	8/17/2017 9:29:40 AM -0500	2/13/2018 8:29:41 AM -0600	9/28/2017 5:26:41 PM -0500	no	no
13147453781779856	.twitter.com	guest_id		/	8/17/2017 9:29:41 AM -0500	8/17/2019 9:29:41 AM -0500	10/2/2017 9:08:23 PM -0500	no	no
13147453811920751	bat.bing.com	MUIDB		/	8/17/2017 9:30:11 AM -0500	8/17/2019 9:30:12 AM -0500	9/28/2017 5:26:41 PM -0500	no	yes
13147453812213595	.spotxchange.com	user-0		/	8/17/2017 9:30:12 AM -0500	8/17/2018 9:30:12 AM -0500	9/28/2017 5:24:49 PM -0500	no	no
13150866340863802	.msn.com	PreferencesMsn		/	9/25/2017 9:25:40 PM -0500	9/25/2018 9:25:40 PM -0500	9/25/2017 9:25:40 PM -0500	no	yes

Fig. 6 Cookies

the standard Chrome browser, there were over 650 cookies that were generated, along with 3 extension cookies and 1 safe browsing cookie. In Incognito we cleared the cookies before we started and there were only 2 cookies which were generated through our data gatherings (Fig. 6).

While browsing through different websites there are sometimes things that happen that we don't realize are going on, and we end up doing something that we didn't mean to do like accidentally download a virus, when we didn't know we downloaded something. Another one of Chrome's privacy settings is to prompt the user where they would like to download the file instead of automatically downloading it into a default folder. This setting prevents a lot of malicious viruses or software from being downloaded onto the computer. Downloads were another big difference that we were able to see between the two browsers. In the normal browser you were able to see every instance of when we downloaded an artifact while we were generating data. It listed data such as: the current file path, where the original target path was when it was downloaded, the URL, the start and end times of the download, total bytes of the artifact, and whether the item was opened or not. This is a lot of user data that about each item that we were able to uncover. Similar to the other artifact gatherings, Chrome's Incognito mode did not gather any of this download data. We cleared the download history before we started, but no data was gathered during our browsing sessions.

Below is a table of the high level artifact data that was captured in both of Chrome's browsers (Table 3):

There are many different types of data that you can gather from the comparison of these two different web browsers, including some data that is not represented in our findings. Many articles read stated that in normal browsing, the username and

Table 3 Chrome Data

Artifacts	Normal browsing	Private browsing
Bookmarks	X	–
cookies	X	–
Extension cookies	X	–
Safe browsing cookies	X	–
History—visits	X	–
History—searches	X	–
History—downloads	X	–
Top sites	X	–
Username	–	–

profile		Object	(3 members)
info_cache		Object	(1 members)
Default		Object	(8 members)
active_time		Real	1.507e+09
avatar_icon		String	chrome://theme/IDR_PROFILE_AVATAR_0
background_apps		Boolean	false
is_using_default_avatar		Boolean	true
is_using_default_name		Boolean	true
managed_user_id		String	
name		String	Person 1
user_name		String	
last_active_profiles		Array	(0 members)
last_used		String	Default

Fig. 7 UserData

passwords would be recoverable [21]. However, in our data gatherings the VMs that contained the data for the username and password got corrupted and were unable to be imaged along with all the other findings. The screenshot below shows where the user data is supposed to be (Fig. 7).

After gathering the data and comparing Google Chrome's data to that of other browsers, Chrome and Mozilla Firefox were very similar in the output of their data. When looking at Google Chrome's Incognito mode and Mozilla Firefox's Private mode both of them were unable to recover much data and ended up being equal in both of the data points that they were both able to recover. Both of these browsers normal modes again were similar in the data that you were able to see. In some of the research read there are more differences that happen under the surface such as "Google Chrome disables all extensions during private browsing mode and Firefox does not" [20]. The few differences were the type of artifacts that were seen due to different data possibly being gathered.

When comparing Google Chrome's Incognito mode and the TOR browser, Chrome had a lot more data that was recoverable then TOR. Part of TOR is based off modified version of Firefox, so it would undoubtably be a bit more secure than Firefox and Chrome. With TOR you were not only unable to see browsing traffic data but also any configuration data, where on Google Chrome you would be able to see the configuration data.

Looking at the findings and data gathered between the Chrome browser's normal and private Incognito browsers there were many differences in what data was stored and what was not. Although Incognito is not an absolute safe solution and it will not block all the traffic that a user generates, it does a lot better job than that of a normal browser. It is a safe bet that if one does not want their data to be stored and tracked using Chrome's Incognito mode is a safe easy way to go.

4.2 Mozilla Firefox

In both the regular browsing and private browsing modes of Mozilla Firefox, browsing actions were performed in a nearly identical manner to reduce any data bias or mismatch which could lead to inaccurate results. Overall, the disk space used by both browsing modes shows initial hints at which type of browsing leaves more artifacts on the system. While the regular browsing machine image had 43.13 GB of unallocated space, the private browsing machine had 43.37 GB of unallocated space—both were 75 GB drives. As expected, any user-initiated downloads remained on the system in the appropriate Downloads and Desktop folders.

In the court of law, forensic browser data can either incriminate or exculpate a user. Based on how and where data is stored for a particular browser, forensic experts can put together a solid assessment and timeline of events that took place based on the browser used. Almost all browser-specific data for Mozilla Firefox is stored in the user's AppData Roaming directory under a Mozilla Firefox Profile. Within the profile folder, various artifacts can be recognized and recovered. The browser artifacts are mostly stored in.json or SQLite formats—both of which can be opened natively in FTK if no other tools are available.

Different addons and extensions which are available in the Firefox web store extend the functionality of the browser and can lead to changes in how browsing and associated data is handled. Knowing which extensions were installed and used can help experts understand how the forensic evidence may have been modified or any new evidence to look for. In FTK, we were able to view information about which browser extensions were downloaded and used, in this case, Privacy Badger and uBlock Origin. While HTTPS Everywhere was installed then uninstalled for both virtual machines, the remnants of the extension did not show up in the Private Browsing artifacts under the Browser Extension Data directory. Advanced users may also decide to modify deeply embedded browser settings and preferences—usually through about:config—which could change behavior of the browser. Examiners can use Prefs.js to view a list of modified settings and preferences for the browser in addition to information about browsing duration and session data (Fig. 8).

```
user_pref("browser.startup.homepage_override.buildID", "20170824053622");
user_pref("browser.startup.homepage_override.mstone", "55.0.3");
user_pref("browser.tabs.remote.autostart.2", true);
user_pref("browser.taskbar.lastgroupid", "E7CF176E110C211B");
user_pref("browser.toolbarbuttons.introduced.pocket-button", true);
user_pref("browser.uiCustomization.state", "{\"placements\":{\"PanelUI-contents\":[\"edit-controls\",\"zc
id1-mnnxcxisbpnsxq_jetpack-browser-action\"],\"dirtyAreaCache\":[\"PersonalToolbar\",\"nav-bar\",\"TabsTc
user_pref("browser.urlbar.lastSuggestionsPromptDate", 20170914);
user_pref("browser.urlbar.timesBeforeHidingSuggestionsHint", 0);
user_pref("datareporting.healthreport.nextDataSubmissionTime", "1386963827105");
user_pref("datareporting.healthreport.service.firstRun", true);
user_pref("datareporting.policy.dataSubmissionPolicyAcceptedVersion", 2);
user_pref("datareporting.policy.dataSubmissionPolicyNotifiedTime", "1471725478298");
user_pref("datareporting.policy.firstRunTime", "1386877427104");
user_pref("datareporting.sessions.current.activeTicks", 4);
user_pref("datareporting.sessions.current.clean", true);
user_pref("datareporting.sessions.current.firstPaint", 1737);
user_pref("datareporting.sessions.current.main", 330);
user_pref("datareporting.sessions.current.sessionRestored", 2015);
user_pref("datareporting.sessions.current.startTime", "1502980073675");
user_pref("datareporting.sessions.current.totalTime", 33);
user_pref("datareporting.sessions.current.index", 9);
user_pref("datareporting.sessions.previous.0", "{\"s\":1386877422244,\"a\":2,\"t\":11,\"c\":true,\"m\":14
user_pref("datareporting.sessions.previous.1", "{\"s\":1386961651758,\"a\":2,\"t\":21,\"c\":true,\"m\":12
user_pref("datareporting.sessions.previous.2", "{\"s\":1386962316502,\"a\":9,\"t\":85,\"c\":true,\"m\":13
user_pref("datareporting.sessions.previous.3", "{\"s\":1471725389088,\"a\":6,\"t\":73,\"c\":true,\"m\":12
user_pref("datareporting.sessions.previous.4", "{\"s\":1471725463522,\"a\":7,\"t\":185,\"c\":true,\"m\":1
user_pref("datareporting.sessions.previous.5", "{\"s\":1471725665591,\"a\":0,\"t\":5,\"c\":true,\"m\":52,
user_pref("datareporting.sessions.previous.6", "{\"s\":1500495496279,\"a\":10,\"t\":243,\"c\":true,\"m\":
user_pref("datareporting.sessions.previous.7", "{\"s\":1500496183385,\"a\":1,\"t\":8,\"c\":true,\"m\":14:
user_pref("datareporting.sessions.previous.8", "{\"s\":1500497605672,\"a\":3,\"t\":14,\"c\":true,\"m\":67
user_pref("dom.apps.reset-permissions", true);
user_pref("dom.ipc.processCount.web", 4);
user_pref("dom.mozApps.used", true);
user_pref("e10s.rollout.cohort", "webextensions-multiBucket4");
user_pref("e10s.rollout.cohortSample", "0.831281");
user_pref("e10s.rollout.cohortSample.multi", "0.325159");
user_pref("experiments.activeExperiment", false);
user_pref("extensions.blocklist.pingCountTotal", 4);
user_pref("extensions.blocklist.pingCountVersion", 4);
user_pref("extensions.databaseSchema", 21);
user_pref("extensions.e10s.rollout.blocklist", "");
user_pref("extensions.e10s.rollout.hasAddon", true);
user_pref("extensions.e10s.rollout.policy", "50allmpc");
user_pref("extensions.e10sBlockedByAddons", false);
user_pref("extensions.e10sMultiBlockedByAddons", false);
user_pref("extensions.getAddons.cache.lastupdate", 1506041390);
user_pref("extensions.getAddons.databaseSchema", 5);
user_pref("extensions.hotfix.lastversion", "20170302.01");
user_pref("extensions.https_everywhere.webextension-migrated", true);
user_pref("extensions.lastAppVersion", "55.0.3");
```

Fig. 8 Prefs.js

In addition to extensions and addons, there are a few other features which improve the browsing experience, but also leave behind interesting forensic information. One of these features is SessionStore.js. By default, Firefox will collect session data from a user so that in the event of some unexpected crash or shutdown, the user will be able to restore the browsing session at a later time as if nothing happened. Private browsing does not keep session data in SessionStore.js. Figure 9 shows some remnants of session data during one of the regular browsing sessions for Gmail and Google Drive.

One common feature many users employ to make web browsing smoother and hassle free is to let the browser remember login usernames and associated passwords. If the saved login feature is used, it would be a good idea to use a master password. A better option would be to use a 3rd party password manager and not store credentials in a browser. The best option would be to memorize all credentials and not store them anywhere. However, if a user chooses to let Firefox remember credentials for various websites, this information is stored in Logins.json. This function was not used in Private Browsing. For Regular Browsing, we did save one login for UTSA single sign-on which was stored in j_username and j_password parameters (Fig. 10).

Logins.json also records dates and times for when the login was 'created' or first stored, the last time it was used, when the password was updated, and how many

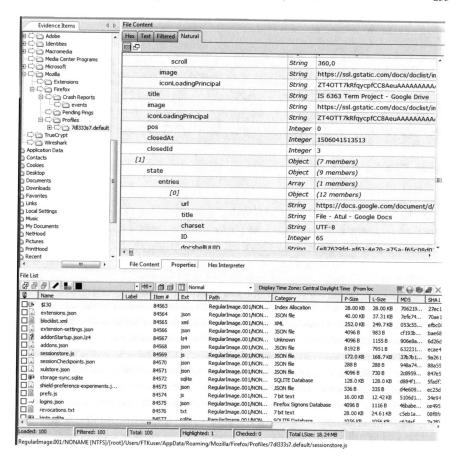

Fig. 9 SessionStore.js

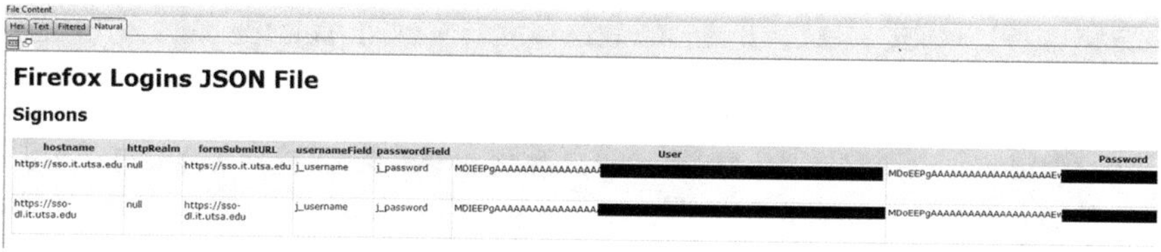

Fig. 10 Logins.json

TABLE moz_formhistory	Search	Show All					
id	fieldname	value		timesUsed	firstUsed	lastUsed	guid
1	id	217700348616695		1	1505435046172000	1505435046172000	vbci3z
2	ev	Microdata		1	1505435046172000	1505435046172000	b/txHl
3	dl	http://lifehacker.com/5940565/why-you-should-start-using-a-vpn-and-how-to-choose-the-...	1	1505435046172000	1505435046172000	pMCg
4	rl	http://gizmodo.com/how-to-hide-your-browsing-history-from-your-snooping-is-1793687193	1	1505435046172000	1505435046172000	3UjhL:
5	if	false		1	1505435046172000	1505435046172000	OG9W
6	ts	1505435046116		1	1505435046172000	1505435046172000	/GeNv
7	cd[Schema.org]	[]		1	1505435046172000	1505435046172000	/Jfforr
8	v	2.7.21		1	1505435046172000	1505435046172000	Y29pv
9	o	28		1	1505435046172000	1505435046172000	wiZwf
10	j_username	jid745		2	15055846534310...	15055868058263000	aXCY/
11	u	google.com		1	1505868225750000	1505868225750000	4mDv

Fig. 11 FormHistory.SQLite

times the login was used. Along with logins, users also have the option to save form data such as usernames, addresses, or phone numbers into web forms. In the case of UTSA login, the username was stored in Formhistory.SQLite under the same j_username parameter (Fig. 11).

While the Login.json values are encrypted, FormHistory.SQLite values are shown in plain-text. FormHistory was not utilized in Private Browsing. One artifact recovered, Signons.SQLite, is no longer used by Firefox after version 32, however it still remains for backwards compatibility. Login.json contains encrypted values, which can be decrypted using key3.db. A forensic investigator could supply these 2 files into FTK PRTK to crack the user passwords which may have been stored in the browser. Another common feature which many people use across various devices—whether or not they have a choice—is to give devices and applications specific permissions to various attributes or data. Just as mobile applications require permissions in order to integrate with other functions, websites can also request permission for things such as access to location, microphones, or webcams. Any website permissions granted through Mozilla Firefox are stored in Permissions.SQLite, but the feature was not explored during this experiment.

When it comes to creating a timeline of events for browsing data, examiners can look to the Places.SQLite file if regular browsing was used. This file can also be used to explore what bookmarks were saved by the user. If private browsing is enabled, then no browsing history nor bookmarks are stored. Internet cookies are another piece of the puzzle which can help forensic examiners determine which sites were visited by a user. If there is a lack of browsing history or lack of cookies stored on the machine, that could be an indication that an attempt was made to erase evidence. In the regular browsing session, 1029 cookies were stored on the virtual machine, while only 2 were stored for private browsing. The 2 cookies stored appear to be default Google cookies, which may be due to a Google search bar function in the browser. As if the Cookies.SQLite file wasn't enough, Firefox used to allow supercookies to be stored via SiteSecurityServiceState.txt (Fig. 12).

While the HSTS information stored in this document can help secure website connections, the HSTS information could cause trouble for a user. Before Firefox version 34, if a cookie was set via regular browsing, the original site along with other sites would be able to view this cookie even if the user switched to private browsing

```
snippets.cdn.mozilla.net:HSTS  5      17431  1537577242991,1,0
www.google-analytics.com:HSTS         4      17430  1516838338739,1,1
drive.google.com:HSTS  0       17431  1537577298325,1,0
self-repair.mozilla.org:HPKP    1      17395
        1503066481360,1,0,WoiWRylOVNa9ihaBciRSC7XHjliYS9VwUGOlud4PB18=r/mlkG3eEpVdm+u/ko/cwxzOMc
startpage.com:HSTS      0      17425  1568658124814,1,1
s5-us2.startpage.com:HSTS      0      17425  1568658148064,1,1
use.typekit.net:HSTS   0      17429  1537487915480,1,1
versioncheck.addons.mozilla.org:HSTS  0       17423  1505353620818,1,0
discovery.addons.mozilla.org:HPKP     0       17430  1511136141708,1,1,WoiWRylOVNa9ihaBciRSC7XHjliYS9V\
e.reddit.com:HSTS      0      17424  1536970826063,1,1
services.addons.mozilla.org:HSTS      7       17431  1537577392087,1,0
stats.g.doubleclick.net:HSTS    3      17430  1516838406227,1,1
query.yahooapis.com:HSTS       0      17429  1537404043400,1,0
www.facebook.com:HSTS   3      17430  1521504049202,1,0
addons.mozilla.org:HPKP         0      17430  1511136312159,1,1,WoiWRylOVNa9ihaBciRSC7XHjliYS9VwUGOlu(
services.addons.mozilla.org:HPKP      6       17431  1511225392088,1,1,WoiWRylOVNa9ihaBciRSC7XHjliYS9V\
versioncheck-bg.addons.mozilla.org:HSTS       3       17431  1506041691007,1,0
hide.me:HSTS  0       17429  1521636227568,1,0
mail.google.com:HSTS  1       17431  1516927920130,1,1
d2yyd1h5u9mauk.cloudfront.net:HSTS    0       17430  1537488049795,1,0
google-analytics.com:HSTS       0      17425  1516472572970,1,1
shavar.services.mozilla.com:HSTS      0       17366  1532031506218,1,1
www.yahoo.com:HSTS  0  17429  1537487871829,1,0
nl.hideproxy.me:HSTS  0       17429  1521636228598,1,0
um.simpli.fi:HSTS      1       17429  1569023882663,1,1
self-repair.mozilla.org:HSTS    2      17395  1534516081360,1,1
www.ixquick.com:HSTS  0       17425  1568658146458,1,1
hangouts.google.com:HSTS       0      17431  1537577511315,1,1
www.mozilla.org:HSTS  0       17395  1534516118646,1,0
www.youtube.com:HSTS   0      17424  1536971624834,1,0
docs.google.com:HSTS  0       17431  1537577468384,1,1
normandy-cloudfront.cdn.mozilla.net:HPKP      4       17429
        1505521171928,1,0,WoiWRylOVNa9ihaBciRSC7XHjliYS9VwUGOlud4PB18=r/mlkG3eEpVdm+u/ko/cwxzOMc
sync.outbrain.com:HSTS         1      17430  1521676877070,1,1
connect.facebook.net:HSTS      2      17430  1521504108825,1,1
www.eff.org:HSTS       0      17430  1537488313953,1,1
facebook.com:HSTS      0      17429  1521503826043,1,0
addons.mozilla.org:HSTS         0      17430  1537488312159,1,0
ssl.google-analytics.com:HSTS  2      17430  1516471013431,1,1
versioncheck-bg.addons.mozilla.org:HPKP       3       17431  1511225391008,1,1,WoiWRylOVNa9ihaBciRSC7XH
apis.google.com:HSTS  2       17431  1537577334345,1,0
youtube.com:HSTS       0      17424  1516321839596,1,1
static.xx.fbcdn.net:HSTS       0      17425  1521138171161,1,1
normandy.cdn.mozilla.net:HPKP  3      17429
```

Fig. 12 Site Security Services

[9]. The data stored in the SiteSecurityServiceState.txt file for private browsing was most likely captured before the private browsing settings were configured and saved since it contained generic Mozilla addon data. HSTS and cookie data in private browsing mode is written to RAM rather than the cookie and HSTS files.

When compared to TOR, Firefox Private Browsing stored more information and data. This is expected since the TOR foundation starts off with a secure Firefox platform with additional security and privacy extensions. In comparison to Google Chrome Incognito, Mozilla Firefox Private Browsing ended up being equal in regard to bookmarks, history, and cookies recovered, where both browsers didn't store any user data or information (Table 4). These 3 items together make up the basis of timeline and activity reconstruction for web browser forensics.

Where TOR stored some execution data in NTUser.dat, Firefox application information could be reviewed in Prefs.js. Prefs.js shows various information about application sessions and when each session was started including the last time the appli-

Table 4 Firefox artifact recovery and location

Mozilla Firefox artifact location & recovery			
Artifact	Regular	Private	Location
Extension data used (block lists, etc.)	X	X[a]	…/browser-extension-data
Crash data	No data[b]	No data[b]	…/crashes
Application health report	No data[b]	No data[b]	…/HealthReport
Application start up and shutdown activity for each session	X	X	…/datareporting/archived/[date]
Extension configuration, setup, associated files, etc.	X	X	…/extensions
Metadata associated with installed extensions	X	X	…/Extensions.json
Default Firefox addon data	X	X	…/Addons.json
Session data	X	–	…/SessionStore.js
Browser preferences	X	X	…/Prefs.js
User-stored logins (remember login for this site?)	X	–	…/Logins.json
Favicons from browsed sites	X	–	…/Favicons.sqlite
Browsing cookies	X	–	…/Cookies.sqlite
Browsing history	X	–	…/Places.sqlite
Key3.db, key to encrypt/decrypt stored user credentials	X	No data[b]	…/Key3.db
Permissions granted to sites (notifications, location, etc.)	No data[b]	No data[b]	…/Permissions.sqlite
User-submitted form history	X	–	…/Formhistory.sqlite

[a]Not complete match between Regular and Private Browsing, but artifacts still recovered
[b]Not applicable based on browsing actions performed

cation was started. Much of the other data mentioned from previous TOR research was memory-only data and a live analysis was not conducted on Firefox.

The discrepancy between Firefox and Chrome regular browsing data volumes could be attributed to either how each browser handles browsing activities, and possible an error made during Chrome data collection where our VMware platform lost one of the snapshot images. All things being equal, default configurations of both Firefox and Chrome browsers should produce similar results for cookies and history data if the browsing actions performed were identical.

4.3 Tor Browser Bundle

The goal of the Tor Browser is to provide its users complete anonymity while brows-ing the internet; this is achieved through a randomized connection of relays in a distributed network making the origin of the user completely undetectable [24]. For this, the Tor Project provides the Tor Browser Bundle (TBB), a complete software package which includes a pre-configured web browser and various extensions, to easily provide a safe and secure way of interacting with resources across the Internet [24].

The Tor browser was specifically chosen as a browser of interest for this project because it utilizes the private browsing modes of Mozilla Firefox when installed using the Tor Browser Bundle (TBB). Although, the Tor project takes it a step further by including several other configurations that prevent any user activity from being written to the disk. To achieve many of its goals, the Tor Project developed the TBB to only store user session data for the duration of the session, this data is then permanently deleted upon session termination [4]. In addition to not writing user session data to the disk, the TBB comes preconfigured with several Firefox extensions which are used to increase security both while on visited sites and to secure the communication between those sites. The use of NoScript [14] allows users to specifically select which sites can execute javascript locally when the site renders. HTTPS Everywhere is a handy extension that forces the use of encryption when using the HTTP/HTTPS protocol; with this extension, all sites are required to default to HTTPS when available [11]. In addition to pre-configured browser extension, the TBB also contains several pluggables that can be used to avoid censorship and proxies; namely—Meek server and ObfsProxy. According to the Tor Project, the Meek Server forces the use of HTTPS and uses a technique known as "domain fronting". This technique hides the fact that you are communicating with a Tor relay [5]. ObfsProxy is another product of the Tor Project which is used by the TBB to alter the appearance of the user's network activity. By encrypting the network traffic and manipulating the underlying protocol, censorship devices that use deep packet inspection are unable to determine that the associated network traffic from the TBB are actually coming from a Tor user [1].

With this knowledge in mind we sought to uncover any remnants of user activity or configuration left behind by the TBB. Utilizing the list of recovered artifacts from the Mozilla Firefox Private Browsing investigation we had a baseline of artifacts we believed would be present. Many of the artifacts recovered from the Firefox Private Browsing sessions were in the Roaming Profile folder for the logged in user. Although the TBB utilizes Firefox, the TBB install process creates a separate directory for storage of the configuration files for the TBB; this is done so that all TBB remnants are contained and can be easily removed from the system. The TBB architecture prevents the browser from writing, or storing (even in temporary files) outside of this directory. While this is a separate directory, several of the configuration files are named the same as those in the Roaming Profile. Runa Sandvik performed a detailed forensic investigation of the TBB in 2013 which focused on version 2.3.25-6 of the

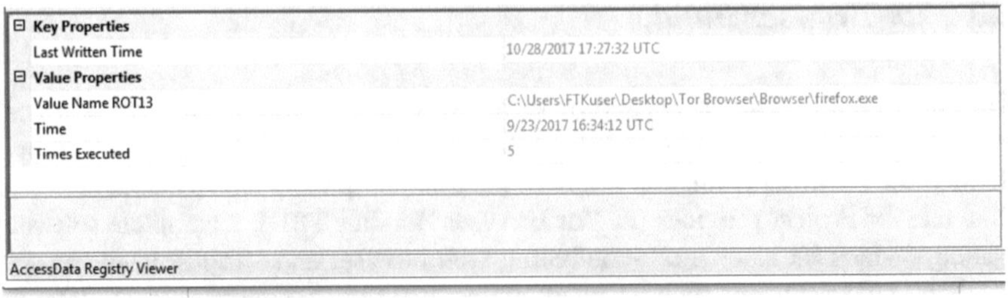

⊟ **Key Properties**	
Last Written Time	10/28/2017 17:27:32 UTC
⊟ **Value Properties**	
Value Name ROT13	C:\Users\FTKuser\Desktop\Tor Browser\Browser\firefox.exe
Time	9/23/2017 16:34:12 UTC
Times Executed	5

AccessData Registry Viewer

Fig. 13 User assist key from NTUSER.dat

TBB [22]. Her analysis recovered several artifacts of interest regarding the usage of the TBB; however, she was unable to uncover user activity which was written to the disk. In our investigation we were able to confirm previous findings regarding the existence of User Assist keys which indicated the TBB had been ran on the investigation media. The userassist key is located in the NTUSER.dat hive of the Registry and indicates the execution path of the program, as well as the number of times the program was executed (Fig. 13).

In addition to the user assist keys, we were able to recover the State file which indicated the last execution time of the TBB. The state file, along with the Torcc, file which indicates the execution path of the TBB, are located in the '…\Data\Tor' directory (Table 5). As with the private browsing mode analysis of Firefox, we were able to recover the installed extensions for the TBB. This information was located in '…\Tor Browser\Browser\profile.default\extensions'. From here we were able to see that both NoScript and HTTPS Everywhere were installed. As mentioned previously, the TBB is a specialized pre-configured version of Mozilla Firefox, and in order for the Tor Project to achieve their goals of complete anonymity there are specialized pieces of software packaged into the TBB. These included pieces of software can be seen in '…\Tor Browser\Browser\Tor Browser\Docs\Sources\versions' as well as '…\Tor Browser\Browser\Tor Browser\Tor\PluggableTransports' which includes the needed components to avoid censorships and proxies. In addition to this information, the specific version of the TBB being used can be found in '…\Tor Browser\Browser\Tor Browser\UpdateInfo\updates\0\update.version' (Fig. 14).

In his research [6], Epifani was able to identify information regarding user visited sites in the PAGEFILE.SYS folder. This information was prefixed with the string "HTTP-memory-only-PB" indicating that it is deleted when the user session is terminated. This leads us to believe that this information is only recoverable during a live system analysis. In our dead system analysis we were unable to recover any of this information from the PAGEFILE.SYS [6].

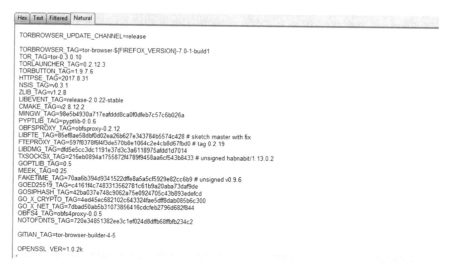

Fig. 14   Built in Tor Browser Bundle Software versions

Table 5   Artifacts recovered from the Tor Browser Bundle

Tor browser bundle (TBB) artifact location & recovery	
Artifact	Location
State file	...\Tor Browser\Data\Tor
Torcc file	...\Tor Browser\Data\Tor
Extensions	...\Tor Browser\Browser\profile.default\extensions
TBB version	...\Tor Browser\Browser\Tor Browser\UpdateInfo\updates\0\update.version
User assist key	%USERPROFILE%\NTUSER.dat
Packaged software versions	...\Tor Browser\Browser\Tor Browser\Docs\Sources\versions
Pluggables	...\Tor Browser\Browser\Tor Browser\Tor\PluggableTransports

5   Conclusion

This research sought to identify the digital artifacts and their locations that could be recovered from various web browsers and web browsing modes. We were able to successfully recover several artifacts of interest, e.g. browser history, cookies, and form autofill information in all public browsing sessions. These findings provided the baseline used for the investigation into several browsers private sessions. In the private browsing sessions of Google Chrome and Mozilla Firefox we were able to recover significantly less than the public browsing sessions, validating several of the claims made by the producers of these programs. When compared to the Tor Browser Bundle, all browsers, even those in private modes, produced more incriminating artifacts. The Tor Browser Bundle excelled in minimizing the amount of information that could be recovered from a user's session. This research provided a thorough comparison of several web browsers and browsing settings.

6 Future Work

Future research in the realm of web browser forensics should continue to analyze which artifacts can be recovered from commonly used web browsers to aid in digital forensic investigations. To get a deeper understanding of the comparison between Chrome, Firefox, and TOR, additional tools and analysis would be needed including live forensic techniques.

With the rise of mobile devices and seamless user experiences, research could focus on mobile web browsers and how browsing sessions among various devices would be handled—switching from desktop to smartphone to tablet using browser login capabilities. These are often marketed to the user as 'pick up where you left off' or 'send to your device' features. In addition, browsing within some native mobile applications might use browser or search APIs, but the data artifacts might be stored and managed by the third-party application rather than the web browser itself.

When it comes to tools and techniques, future research should evaluate available tools which can assist investigators in obtaining accurate information to support a case in a legal manner. Some of these tools might include data or target specific functionality such as SQLite extraction, live memory analysis, or tools focused directly on web browser forensics. Our research was limited by the tools and licenses available to us in a student lab environment. Additionally, research should continue to work on current software versions; Mozilla released an updated version of Firefox—Quantum—shortly after our research concluded.

References

1. Arma.: Obfsproxy: The Next Step in the Censorship Arms Race (2012) Retrieved 19 Nov 2017, from https://blog.torproject.org/obfsproxy-next-step-censorship-arms-race
2. Bagley, R., Ferguson, R.I., Leimich, P.: Firefox Browser Forensic Analysis via Recovery of SQLite Artefacts from Unallocated Space (2012)
3. Boggs, R.J., Fenger, T., Sammons, J., Winkler, D.: Online anonymity: forensic analysis of the tor browser bundle (2017)
4. Clark, E., Koppen, G., Murdoch, S., Perry, M.: The Design and Implementation of the Tor Browser [DRAFT] (2017). Retrieved 19 Nov 2017, from https://www.torproject.org/projects/torbrowser/design/#disk-avoidance
5. Dcf.: How to Use the "meek" Pluggable Transport. Retrieved 20 Nov 2017, from https://blog.torproject.org/how-use-meek-pluggable-transport (2014)
6. Epifani, M.: TOR FORENSICS ON WINDOWS OS (2014). Retrieved 16 Nov 2017, from https://digital-forensics.sans.org/summit-archives/dfirprague14/Tor_Forensics_On_Windows_OS_Mattia_Epifani.pdf
7. Gabet, R.M.: A Comparative Forensic Analysis of Privacy Enhanced Web Browsers (Doctoral dissertation, Purdue University) (2016)

8. Ghafarian, A., Seno, S.A.H.: Analysis of privacy of private browsing mode through memory forensics. Int. J. Comput. Appl. **132**(16) (2015)

9. Goodin, D.: Browsing in Privacy Mode? Super Cookies can Track you Anyway. Ars Technica (2015)

10. Greenwald, G., MacAskill, E.: NSA Prism program taps into user data of Apple, Google and others. Guardian **7**(6), 1–43 (2013)

11. HTTPS Everywhere. (n.d.). Retrieved November 16, 2017, from https://www.eff.org, https://www.everywhere

12. Keller, K.: The Tor browser: A forensic investigation study (Doctoral dissertation, Utica College) (2016, December)

13. Mahaju, S., Atkison, T.: Evaluation of Firefox Browser Forensics Tools. In Proceedings of the SouthEast Conference, pp. 5–12. ACM (2017)

14. Maone, G. (n.d.). NoScript Security Suite by Giorgio Maone. Retrieved November 20, 2017, from https://addons.mozilla.org/en-US/firefox/addon/noscript/

15. Mathews, L.: What Is Private Browsing And Why Should You Use it? Forbes (2017)

16. Mozilla.: A Better, Faster, Private Browser for Today|Firefox. Mozilla (2017)

17. Nalawade, A., Bharne, S., Mane, V.: Forensic analysis and evidence collection for web browser activity. In: International Conference on Automatic Control and Dynamic Optimization Techniques (ICACDOT), pp. 518–522. IEEE (2016, September)

18. Noorulla, E.S.: Web Browser Private Mode Forensics Analysis. Rochester Institute of Technology (2014)

19. Oh, J., Lee, S., Lee, S.: Advanced evidence collection and analysis of web browser activity. Digital Invest. **8**(S62), S70 (2011)

20. Ohana, D.J., Shashidhar, N.: Do private and portable web browsers leave incriminating evidence?: a forensic analysis of residual artifacts from private and portable web browsing sessions. EURASIP J. Inf. Secur. **2013**(1), 6 (2013)

21. Rathod, D.: Web browser forensics: Google Chrome. Int. J. **8**(7) (2017)

22. Sandvik, R.A.: Forensic Analysis of the Tor Browser Bundle on OS X, Linux, and Windows (2013). Retrieved from https://research.torproject.org/techreports/tbb-forensic-analysis-2013–06-28.pdf

23. Shafqat, N.: Forensic investigation of user's web activity on Google Chrome using various forensic tools. Int. J. Comput. Sci. Netw. Secur. **16**(9), 123–132 (2016)

24. Tor: Overview. (n.d.). Retrieved 14 Nov 2017, from https://www.torproject.org/about/overview.html.en

25. Tsalis, N., Mylonas, A., Nisioti, A., Gritzalis, D., Katos, V.: Exploring the protection of private browsing in desktop browsers. Comput. Secur. **67**, 181–197 (2017)

26. Warren, A.: Tor Browser Artifacts in Windows 10 (2017)

Forensic Analysis of Microblogging Sites Using Pinterest and Tumblr as Case Study

Briana McFadden, Elamathi Balasubramani and Wakama E. Miebaka

Abstract The growth in the popularity of microblogging and online social network (OSN) websites such as Facebook, Twitter, Tumblr, Pinterest, etc. in this modern age progresses relentlessly. Accompanying this growth in popularity is the increase in the use of these websites by individuals, companies, and even industries at large. Aside from social networking, microblogging and OSN websites have been exploited by many profitable companies as a medium for advertisement of their products, information dissemination and building connections. This increasing usage for various purposes has also opened doors to a wide range of criminal and malicious acts. This chapter studies two particular microblogging sites, Pinterest and Tumblr, both of which have gained great popularity today. Specifically, we seek to contribute to the forensics field by finding the artifacts associated with these websites that would be of interest should the need arise, as in criminal and civil cases, by replicating the actions of a standard user while using these sites in the Internet Explorer, Firefox and Chrome Browsers, and then, analyzing this self-created evidence to find such artifacts. Also, the paper exposes and exploits most of the top forensic tools employed today in the investigation of cyber-criminal cases, while making use of forensically sound methodologies in our analysis.

1 Introduction

Micro-blogging or blogging forms a major part of information dissemination today and has won over a lot of people as a way to gain information, communicate, and connect with others in the world. People blog (make posts online) to express their opinions on different subject matters and to have a wide conversation with a large audience [1]. Different organizations and companies have employed it as a means of marketing and advertising their products and services. Some popular micro-blog or blog networks include Twitter, Tumblr, Pinterest, etc. But, as with everything that has

B. McFadden · E. Balasubramani (✉) · W. E. Miebaka
Department of Information System and Cyber Security,
University of Texas at San Antonio, San Antonio, TX, USA
e-mail: belamathi@gmail.com

© Springer Nature Switzerland AG 2020
X. Zhang and K.-K. R. Choo (eds.), *Digital Forensic Education*,
Studies in Big Data 61, https://doi.org/10.1007/978-3-030-23547-5_13

been created, there comes corruption and abuse by ill-intended people. Unfortunately, micro-blogging or blogging is no different.

Blogging abuse or misuse appears in many different shapes when it comes to governmental and private organizations. There have been different cases of misconducts that have happened through blogging. Common among them is cyberstalking and cyberbullying which have inflicted harm and has even led to the death of a person [1]. Other forms of misuse and abuse include companies or individuals gaining access to and then leaking trade secrets of rivalry companies. There have also been reported cases of individuals trying to influence the decisions of jurors [1]. These scenarios mentioned only scratch the service of micro-blogging and blogging misuse and abuse. Due to the intrinsic behavior of individuals to abuse and misuse micro-blogging and blogging website, the evidentiary and useful information that can be derived from these blogs hold great value to an investigation process. Artifacts derived may be directly admissible as circumstantial evidence in court, and the psychology or mental state of a person may be studied by the kind of posts or blogging that he performs.

It would come as no surprise that individuals try to cover their tracks when they perpetrate these crimes by taking some anti-forensics measures to remain anonymous to their audience. How forensic investigators approach such cases represents a delicate issue due to the complexity of online social networking as a whole. Unfortunately, there is not a lot of ongoing forensics research in realm of micro-blogging and blogging. We seek to explore this area, specifically to find out what forensic artifacts may be useful in an investigation process using forensically sound approaches or methods that have been developed over the years.

2   Related Work

Past forensic endeavors undertaken by some researchers in online social networking may be grouped into online OSN forensic investigation principles and methodologies and browser analysis.

2.1   Online Social Networking (OSN) Forensic Investigation Principles and Methodologies

Forensics, being a scientific endeavor, cannot be successfully accomplished without a defined set of methods or models through which its processes must follow. The methodology problem in forensics has been ongoing and tackled by bright minds over many years. Most models that have been speculated tend to follow same approach and are quite similar to each other [2]. The fast rising nature of social networking, its ubiquity and intricate nature poses a new type of problem that demands new defined methodologies when being investigated [3], as discussed below.

Why is there a need for a specific methodology for forensic investigation of online social networking? Without much thought, this does not pose a problem, but this is quickly refuted by the numerous challenges associated with social networking. Unfortunately, previous models of forensics did not fit the profile of online social networking [2, 4]. Madjid et al. in [2] identifies the challenge of obtaining authorization for evidence acquisition from online social network servers, thus the need for a *preparation/preliminary* phase. Dr. Mark et al. in [3] also points out the need for a forensic investigator of OSN to have a comprehensive understanding of the social network being investigated. Social networks contain data evidence that can be easily manipulated or changed online, e.g. the change of emoticons, pictures and texts, before it is secured by the investigator. This poses a time-constraint problem to the forensic methodology. Dr. Mark suggests that snapshots or video-recording with a camera from the user or investigator can be very useful in the investigations. Jang, Yu-Jong, and Jin Kwak identifies the problem with accessing OSNs from a device, it leaves only weak evidence information such as usage log records on the device, but that useful evidential information can be found in the live-memory in the part that corresponds to the application or browser used in accessing the OSN. They also mention the issue of inaccessibility to data found in private accounts, inefficiency of web browser records to be used as digital evidence, and the modification or deletion of data within a mobile device.

With all of these challenges in sight, it is worth noting, as mentioned in [3] that the best OSN evidence is most likely obtained from the OSN servers with an on-time authorization granted. Jang, Yu-Jong, and Jin Kwak takes this a bit further by the suggestion of cutting-off the network connection and suspension of the suspect's account to prevent further modification by any user. However, when data acquisition from the servers is not possible, we are left with the client's device to obtain our evidence [3]. This is not to be underestimated as the evidences obtained from client devices have proved to be very useful over the years. The images, videos, conversation or message texts, when analyzed, can be very useful to the investigation process. As mentioned before, live-memory analysis on the client device can reveal artifacts such as the profiles visited by the user, the time of access to the social network and message texts transacted between the user and other individuals.

To address these challenges comprehensively, most of the aforementioned authors have developed models specifically for the investigation of OSN. Apparently, most of them tend to have very similar features with the initial stage of *preparation/preliminary* seeming to be consistent across the developed models. This is very understandable, having considered the challenges mentioned above and the complexity associated with OSN, the need for a timely preparation is needed. Madjid et al. went on to develop an application prototype that implements a large part their proposed model when fed in data to expedite or automate the investigation process. Other notable models mentioned included those written by Philip D. Dixon, Lee et al. and the United States Department of Justice. Philip D. Dixon's model incorporated five main goals, preservation, identification, extraction, documentation and interpretation of computer data. Lee et al. wrote a model named *The Scientific Crime Scene Investigation Model* which had four steps: recognition, identification, individualization, and reconstruction. The United State Department of Justice created a model that consisted of four steps as well which is: collection, examination, analysis and reporting. Overall, there are a good deal of social media application and online social network investigation principles, models, and methodologies. While this is a good thing, this is also a challenge as there is no standardization when it comes to digital forensics investigations.

2.2 Browser Analysis

This research is mostly concerned with accessing microblogging/blogging sites using *web browsers*. Because of this, it makes absolute sense to consider the previous works that have analyzed web browsers access to OSNs for forensic purposes. It has been identified that evidence relating to a case can be found on the social media network and that the best way to obtain this evidence would be through the consent of the user being investigated. With no consent of the user, to garner evidence from the social media network, it would take the use of law enforcement that would need to obtain and submit orders such as preservation orders, search warrants, court orders, and/or subpoenas.

The browser is actually a great source of potential evidence. Things that can be found through the browser include search history, browser history, sites visited, stored login and password information, and the cache which holds copies of previously visited websites. Metadata of social media posts can actually be found and used as evidence through investigating the browser. Metadata can be used to answer the main questions, such as, who, what, when, how, where and why, of a digital forensics investigation [5].

Moreover, different methodologies have been proposed for browser analysis. Most research studies focused on obtaining data from a single browser at a time, but in reality, a single user can use multiple web browsers to perform their internet searches at the same time [6]. Junghoon Oh and his associates propose an integrated analysis of evidence collected from the log files, where all browsers are analyzed at the same time and in a combined time-frame taking into account the different time format of the browsers [6]. In addition, they propose a way to retrieve the search keywords used from the URL's visited. Murilo Tito Pereira [7] of Brazilian Federal Police proposes an algorithm to recover Firefox history records from unallocated space. This algorithm is specifically designed to recover entries from the *moz_places* table of Firefox [7]. Rahman [8] and his associate focus on live forensics by profiling different applications. They identify the number of processes, services and modules that are loaded when launching an application program. Investigators can collect not only the information about live processes but also about the terminated and cache processes using this profile of an application [8].

2.3 Artifacts Found in Existing Literature

See Table 1.

2.4 Summary of Tools Used in Existing Literature

See Table 2.

3 Browsers Explored

- Internet Explorer v11
- Firefox v56
- Chrome v61

4 Tools Used in Analysis

See Table 3.

5 Flow Chart

See Fig. 1.

6 Evidence Creation

The Evidence Creation stage constitutes those activities that are carried out to replicate or represent the behavior of a standard user (a user with no criminal or ill intentions) when using the various microblogging sites through different web browsers.

Before acting out the activities of a standard user and those pertinent to this
analysis the browser's option settings have to be configured, such as enabling cookies,
time it takes to recycle the browser history, remembering passwords and username,
etc. This determines to a large extent how much data will be created and stored in
the user's computer or device. Also, an awareness of the current settings for these
browser options would help us understand and accurately interpret what data should
or should not be stored by the browsers which in turn, would lead to more accurate

Table 1 Artifacts found in existing literature

Refs. No	IE	Chrome	Firefox	Safari	Opera
Seigfried-Sellar and Leshney [10]	Browser history, searches, sites visited, stored login and password information, cache, social media posts metadata		Browser history, searches, sites visited, stored login and password information, cache, social media posts metadata		
Rathod [11]		Default artifacts locations: history, cookies, login data, topsides, shortcuts, user profile, prefetch file and RAM dump			
Pereira [7]			History, cookies and user data		
Oh et al. [6]	Log files				
Jones [12]	Identifies file size, hash table, directories, URL				
Ohana and Shashidhar [13]	Browsing history usernames email accounts images	Browsing history images	Browsing history images		
Akbal et al. [14]	Cookies	Reference files	Places.sqlite		
Said et al. [15]	In Physical memory				
Rahman and Khan [8]		Profile of chrome			
Jang and Kwak [16]	In-memory: Facebook profiles visited, timeline of conversations or messages, type OSN used				

Table 2 Tools used in existing literature

Forensics	Tools used						
	FTK Imager and FTK	Encase	DD	HxD	Belkasoft	Chrome/IE/ Firefox cache viewers	Hex editor
Social media applications	Facebook, Skype [17]		Facebook, Skype [17]	Facebook, Skype [17]		Facebook, Skype [17]	
Browser analysis	Firefox, IE, Chrome, Opera, Safari [14, 15]	Firefox, IE, Chrome, Opera, Safari [14, 15]		Chrome [11]	Chrome [11]	Firefox, IE [13, 15, 17]	IE, Firefox, Chrome [12, 15]

Table 3 Tools used in analysis

Tools	Usage
FTK imager v4.1.0.12	To create forensic image of VMDK files
AccessData forensic toolkit v6.2.0.1026	Processing and analysis of imaged evidence
SQlite viewer with google drive v1.0.2.1	To view the SQLite database files of Firefox
PRTK v6.2.0.1026	To recover user id and password

results. However, for the sake of this analysis, all options were left in their default settings. The table below (Table 4) is an overview of the main settings that may affect this considerably.

Every activity carried out during the evidence creation, bearing in mind the *Locard's exchange principle*, was done to explore and replicate both the natural behavior taken by a standard user and the evasive actions or measures that may be taken by an ill-intended individual to cover his or her tracks while using either of the blogging sites, Tumblr and Pinterest. Each of these of activities was repeated for each of the browsers (Internet Explorer, Firefox and Chrome).

After each day of evidence creation, snapshots were taken of the live computer, i.e., before shutting down or closing the browsers, to preserve the memory or volatile data for further analysis, and also if we should have any reason to have to revert to any of the snapshots taken in previous days for investigation.

Below is a breakdown of some of the activities carried out and the rationale behind them (Table 5).

The evidence creation extended across multiple non-consecutive days and came up to the number of days and hours as shown below in the statistics table below (Table 6).

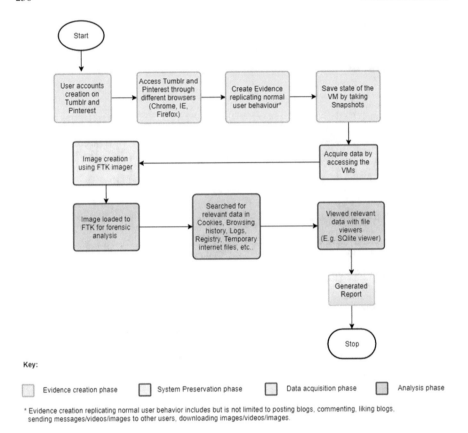

Fig. 1 Flowchart of the forensic process performed

Table 4 Overview of the browser settings

Default options	Internet explorer v11	Chrome v61	Firefox v56
Enable cookie	✓	✓	✓
Remember form data	✓	✓	✓
History recycle time	20 days	—	—
Download folder	C:/User/[user]/Downloads/	C:/User/[user]/Downloads/	C:/User/[user]/Downloads/

Table 5 Activities performed during evidence creation

Activities	Tumblr	Pinterest	Rationale
Create an account	✓	✓	Dummy accounts were created for the sole purpose of this exercise. No real email accounts were used
Login	✓	✓	The time of login, as well as, login credentials if captured may prove valuable during the investigation process
Create dashboards		✓	Pinterest allows the creation of dashboards were pins are saved. Its direct usefulness is not known for now, but it's time of creation as well as its content may be useful also
Upload and save pins to dashboard		✓	As opposed to saving images to computer, Pinterest allows for images and videos (pins) to be saved to dashboards. This represents a virtual storage location that must be investigated for evidence. Also the kind of pins uploaded or saved tells a lot about an individual
Comment on posts	✓	✓	Comments comprising both text and emoticons, if discovered can reveal a lot into a case and the time when comment was made can give insight into timeline of event
Play audio and videos	✓	✓	The kind of videos and audio played by an individual may expose the psychological state of mind of the individual
Make some searches	✓	✓	Although not as important as the rest, It may contain some useful information
Uploaded and download images to computer	✓	✓	Images saved to computer and their contents may definitely constitute direct and incontrovertible evidence in a case

(continued)

Table 5 (continued)

Activities	Tumblr	Pinterest	Rationale
Message different parties	✓	✓	Messages sent or received between the dummy accounts created comprise images, text and emoticons. The usefulness of the content of the messages sent and time of conversation cannot be overemphasized and may also serve as direct or incontrovertible evidence too
Clear browser cache and delete downloaded images	✓	✓	It's not uncommon for individuals, for whatever personal reasons, to erase their browsing history or cache and even delete potential evidence in form of images and videos. We seek to find out what evidence or data can be recovered and if the timeline of deletion can be established
Follow and unfollow and visit other users blog	✓	✓	A user's network of followers or friends, as well as blog sites visited and time of visit may open doors in determining the scope of an event and the individuals involved in a case
Make posts	✓	✓	Posts made constitute images, videos, texts and hash tags. The time in which posts were made and its content may reveal timeline of event and their contents may stand as direct evidence

Table 6 Evidence creation statistics

Browser	Days of activities (days)	Hours of activities
Internet Explorer	10	35 h 27 min
Firefox	10	10 h 58 min
Chrome	11	10 h 17 min

7 Preservation and Acquisition of Evidence

After the evidence creation, a forensic duplicate was taken of each of the memory (.vmem) and virtual disk (.vmdk) files, using AccessData FTK imager, before acquiring the created evidence for analysis. Moreover, before each imaging process, hashes (MD5 and SHA) were computed for the files and stored for integrity purposes, and after each imaging process, the hashes of the forensic duplicates are automatically validated by FTK imager with that of .vmdk and .vmem files, ensuring that images are exactly the forensic duplicates of the .vmdk file.

Every snapshot that was taken constitutes changes that were made after the previous ones, and so, to expedite the analysis process, in lieu of analyzing each snapshot separately, the .vmdk files of each snapshot were aggregated to a single .vmdk file using vmware-vdiskmanager, which is a concomitant application with VMware workstation, for analysis using AccessData Forensic Toolkit. Aggregating the .vmdk files of each snapshot into one is completely valid as we are interested in the artifacts that may have been created in each snapshot all together, and not necessarily, the changes after each snapshot that was taken. The hash of the singular .vmdk file was then computed and stored. The .vmdk file, also, was then imaged and mounted on the Forensic Toolkit for analysis.

8 Analysis Using AccessData Forensic Toolkit

Analysis of the evidence files were carried out in these three main areas.

8.1 Decrypting Stored Form Data

8.1.1 Internet Explorer v11

With the advent of Windows Vista and Windows7, there is a new robust encryption algorithm called DPAPI (Data Protection Application Programming Interface) that does not store the recovery key in the local machine, designed to make the stored or protected form data more secure than in previous versions. Should the browser be set to remember form data (data entered by the user in pages embedded form) this form data would be stored in a sub key in the NTUSER.DAT file called *intelliforms,* that has two sub keys—storage 1 (autocomplete form data) and storage 2 (passwords).

The recovery key is not stored in the local machine, however, there is a master key derived from the user's windows login password from which a temporary symmetric session recovery key is generated to encrypt the form data and then stored in the aforementioned storage locations. To decrypt the form data, we need:

- User's login password—The user's password if not known, it can be derived by supplying the SAM and SYSTEM file of the user computer to AccessData PRTK, which would carry-out a password attack on the stored password hashes in the SAM file.
- Protect folder—The protect folder in the Windows registry contains information needed to generate the Master key.
- URL of website used in auto-complete process—The URL of the websites of concern of which you need the form data of. Supplying the location of the stored URLs, such as the IE history location should be sufficient here.
- NTUSER.DAT file containing the encrypted data in the aforementioned storage locations.

With all these files available, AccessData PRTK automates the decryption process by providing an interface where you can simply input these files locations, and then outputs the decrypted form data after the decryption process.

These files were exported from our evidence in FTK. However, in an attempt to decrypt this form data, we were quickly met with no success, as PRTK could not identify any form data in the supplied NTUSER.DAT file. Investigating PRTK could not find any form data, it was found that neither of the storage sub keys (storage 1 and storage 2) was found in the intelliforms key, although it is confirmed, that the browser options was set to remember user form data. Also, the default option was set to "Ask user" before storing form data. This is why the form data were not eventually stored in the local computer or the storage sub keys were not present, as we we would not be asked to save our search queries or other form data in either of Pinterest or Tumblr websites once logged in. Also worth noting, is that we weren't also given the option of whether to save our login username and passwords.

8.1.2   Firefox v56

The password that was stored in the Firefox browser for Tumblr and Pinterest can be found in Logins.json in encrypted format and the key for decrypting the username and password is located in the Key3.db. The username and password was recovered using PRTK, there was no master password set for the Logins.json file in our case so, the username and password were recovered by loading the Logins.json file and providing the key file, Key3.db in PRTK (see Fig. 1 Appendix A).

8.1.3   Chrome

The login data database stores user's login data. Chrome users have two options when it prompts them to save a password for future logins, which are "save password" or "never save password." Depending on which option a user selects determines how much of the data is going to be cached into the logins data database. The database includes many different field values, such as, URL or domain, type of data required, plain text of required data, encrypted password value, and creation date of data. Chrome utilizes the Windows DPAPI cryptographic system to encrypt and decrypt values cached in the Login Data Index and a temporary "session recovery key" is generated to display the plain text of the passwords to the user. One can look at the index inside of the Login Data database file. Here username values, passwords, and other fields mentioned are displayed.

8.2   Investigating Well-Known Locations

8.2.1   Internet Explorer v11

The common web artifacts locations associated with IE are the history, temporary internet files/Web cache, and Cookie locations, the locations of which are easily pointed out by FTK in the "internet and Chat files" tab. Prior to investigating the files in these location, a customized filter was created based on the timeline of the browser activity, using the time of creation and modification of the files as metrics of the timeline. This expedites the analysis process by helping us narrow our scope to only data created or modified during the evidence creation exercise. Analysis or investigation carried out in these locations resulted in the various web artifacts (shown in Table 7). Note also that the locations of the common web artifacts are slightly different from those associated with previous versions of Internet Explorer.

8.2.2   Firefox v56

The common artifact locations for Firefox were analyzed; the location of the artifacts and its details is presented in the following table. Firefox v56 stores the saved user Id and password for a site when the user selects save password and it is no longer stored in the SQLite database file, signons.sqlite but is stored in a is in file, logins.json [9]. As in IE analysis, FTK listed most of the known artifacts in the "internet and Chat files" tab under Mozilla/Firefox. Table 8 lists these well-known artifact locations and provides its details.

Table 7 IE artifacts from well-known locations

	S. No	Artifact	Location	Description
Pinterest		*History*		
	1	Time of login time of logout created pins topics username Pins opened dashboard owned or created	[root directory]/Users/[user]/AppData/Local/Microsoft/Windows/WebCache/WebCacheV01.dat	History contains the named artifacts, explicitly or implicitly, in the url of the web pages visited, but contrary to Tumblr, doesn't show the searches (See Appendix B8)
		Cookies		
	1	–	[root directory]/Users/[user]/AppData/Roaming/Microsoft/Windows/Cookies/Low	It contains Pinterest generated first-party cookies containing some values. However, the purpose of this cookies nor their values are not known (See Appendix B5-6)
		Temporary Internet Files		
	1	Time of logout	[root directory]/Users/[user]/AppData/Local/Microsoft/Windows/Temporary Internet Files/Low/Content.IE5/CF3CV8R6/7339RI1B.htm	It is an htm file that was created in the Temporary Internet directory that corresponds to the time the user logged out (See Appendix B2)
		DOMStore File		
	2	Time of log in	[root directory]/Users/[user]/AppData/LocalLow/Microsoft/Internet Explorer/DOMStore/B7W1L920/www.pinterest[1].xml	The xml file contains a Pinterest login page; whose creation time corresponds to time of login (See Appendix B1)
Tumblr		*History*		
	1	Searches made dashboard of other users pages visited	[root directory]/Users/[user]/AppData/Local/Microsoft/Windows/WebCache/WebCacheV01.dat	The IE History shows every search that is made on Tumblr and the dashboard of every user in the url of the pages visited (See Appendix B16)
		Cookies		

(continued)

Table 7 (continued)

S. No	Artifact	Location	Description
1	A page visited	[root directory]/Users/[user]/AppData/Roaming/Microsoft/Windows/Cookies/Low/FKGYVMFE.txt	A naturally deleted cookie showing the url of a page that was visited. This cookie is quite different from the others (See Appendix B14)
2	—	[root directory]/Users/[user]/AppData/Roaming/Microsoft/Windows/Cookies/Low/UX8KRK76.txt	It is a Tumblr generated cookie containing some values. However, the purpose of this cookie nor the values is not known (See Appendix B15)
Temporary Internet Files			
1	Time of logout	[root directory]/Users/[user]/AppData/Local/Microsoft/Windows/Temporary Internet Files/Low/Content.IE5/D15UCQYW/logout[1].htm	An unallocated htm file showing a logging out activity from Tumblr (See Appendix B10)
2	Uploaded video	[root directory]/Users/[user]/AppData/Local/Microsoft/Windows/Temporary Internet Files/Low/Content.IE5/FH85CC7D/tumblr_owv7i7kPl11wbsfa9_480[1].mp4	An mp4 of a wildlife video that was uploaded to Tumblr from the local drive of the computer
3	Posted video	[root directory]/Users/[user]/AppData/Local/Microsoft/Windows/Temporary Internet Files/Low/Content.IE5/FH85CC7D/tumblr_owv7i7kPl11wbsfa9_480[2].mp4	An mp4 of a wildlife video that was posted
4	Videos embedded on web page visited thumbnails	[root directory]/Users/[user]/AppData/Local/Microsoft/Windows/Temporary Internet Files	These videos, played or not played, images embedded in web page visited are saved as temporary internet files
DOMStore File			
1	Time of logout	[root directory]/Users/[user]/AppData/LocalLow/Microsoft/Internet Explorer/DOMStore/3CFWU0LY/www.tumblr[1].xml	The www.tumblr.xml [5] file shows the date and time of log out (See Appendix B9)

Table 8 Firefox artifacts from well-known locations

Firefox	S. No	Artifact	Path	Details
	1	Pref.js	Firefox_Image.001/Partition 1/NONAME [NTFS]/[root]/Users/FTKuser/AppData/Roaming/ Mozilla/Firefox/Profiles/7dl333s7.default/prefs.js	Gives the user preference on the browser. Sanitizing when shutting down and prompt on sanitize are not set to true hence the user details are not deleted when closing the browser
	2	Logins.json	Firefox_Image.001/Partition 1/NONAME [NTFS]/[root]/Users/FTKuser/AppData/Roaming/ Mozilla/Firefox/Profiles/7dl333s7.default/logins.json	Stores the user password for the site. Inspection on logins.json shows that user saved password for One Pinterest account and two Tumblr account
	3	Key3.db	Firefox_Image.001/Partition 1/NONAME [NTFS]/[root]/Users/FTKuser/AppData/Roaming/ Mozilla/Firefox/Profiles/7dl333s7.default/key3.db	Contains the key for unlocking Logins.json
	4	cookies.sqlite	Firefox_Image.001/Partition 1/NONAME [NTFS]/[root]/Users/FTKuser/AppData/Roaming/ Mozilla/Firefox/Profiles/7dl333s7.default/cookies.sqlite	Recovered cookies and compared it to the logs created during the evidence creation process
	5	formhistory.sqlite	Firefox_Image.001/Partition 1/NONAME [NTFS]/[root]/Users/FTKuser/AppData/Roaming/Mozilla /Firefox/Profiles/7dl333s7.default/formhistory.sqlite	Recovered forms search and compared it to the logs created during evidence creation process
	6	Places.sqlite	Firefox_Image.001/Partition 1/NONAME [NTFS]/[root]/Users/FTKuser/AppData/Roaming/ Mozilla/Firefox/Profiles/7dl333s7.default/places.sqlite	Contains the links of all sites visited
	8	favicons.sqlite	Firefox_Image.001/Partition 1/NONAME [NTFS]/[root]/Users/FTKuser/AppData/Roaming/ Mozilla/Firefox/Profiles/7dl333s7.default/favicons.sqlite	Lists favorite icons based on the sites frequently visited
	9	Downloads	Firefox_Image.001/Partition 1/NONAME [NTFS]/[root]/Users/FTKuser/Downloads/	Downloaded images, videos and GIFs were recovered

Table 9 Chrome artifacts from well-known locations

Google Chrome artifacts			
S. No	Artifact	Path	Details
1	Downloads	PinteresttumblrImage.001/Partition 1/NONAME [NTFS]/[root]/Users/FTKuser/Downloads	Images and videos that were downloaded during social media sessions
2	Browsing history	PinteresttumblrImage.001/Partition 1/NONAME [NTFS]/[root]/Users/FTKuser/AppData/Local/ Google/Chrome/User Data/Default/local storage	All websites visited through browser shows login and logoff times from social media websites investigated

8.2.3 Chrome v61

See Table 9.

8.3 Indexed and Live String Searching

Having checked the known locations for possible artifacts, indexed and live searches for logged items and keywords of interest particularly associated with Pinterest and Tumblr is carried out for other possible locations that data from the user browsing activities may be created and stored.

Apart from searching for the logged items and keywords of interest, using the developer tools of the browsers, back-end data that could be searched for can be discovered. Also, drilling into the back-end of the browsers may help provide context to and make sense of some of the data, e.g. JSON and Javascript files, discovered during the analysis. The tables below (Tables 10, 11, 12 and 13) show the artifacts for each of the browsers that were found from search process:

8.3.1 Internet Explorer v11

See Table 10.

8.3.2 Firefox v56

See Table 11.

Table 10 IE—indexed and live string search results

Pinterest	S. No	Artifacts	Location	Details
	1	Created dashboard	[NTFS]/[MetaCarve]/61002/base06ec.kdc	A metacarved.kdc file showing the name of a created dashboard (See Appendix B3)
	2	Visited pages	[root directory]/Users/[user]/NTUSER.DAT	The NTUSER.DAT contains urls of pages visited, some which that don't appear in the *type urls* key when it is loaded in a registry viewer (See Appendix B4)
Tumblr	1	Messaging between two of the created dummy accounts	[NTFS]/[unallocated space]/03674650/07184694	The unallocated space shows messaging taken place between the names of two participants (See Appendix B4)
	2	Blog that was visited	[root directory]/Users/[user]/AppData/Local/Microsoft/Internet Explorer/Recovery/Last Active/{FE47E47C-B44B-11E7-98EF-000C29A97838}.dat	A.dat file showing a user blog that was visited. Inside it is the url of the visited blog (See Appendix B11)

8.3.3 Chrome v61

See Table 13.

9 Summary on Artifacts Found

Multiple artifacts have been found in all three browsers through the analysis as shown in the tables above, most of which came from the well-known locations for artifacts associated with the browsers. A large part of these artifacts identified have met with our earlier agenda for the artifacts we had set out to look for, while some of the artifacts such as comments made on posts, emoticons used, hashtags used, messages sent to and from users, follow and unfollowing a blog, etc. remained unrevealed throughout the analysis.

However, some the artifacts that were interesting and stood out amongst the rest to us were those that showed messaging or conversation taking place between participants in unallocated and slack spaces of the local disk and in the browser cache. This might be due to the operating system, Windows 7 in our case, filling up slack spaces

Table 11 Firefox-indexed search results for Pinterest

Pinterest	S. No	Artifacts	Location	Details
	1	Thumbnails	Firefox_Image.001/Partition 1/NONAME [NTFS]/[root]/Users/FTKuser/AppData/Local/Mozilla/Firefox/Profiles/7dl333s7.default/thumbnails/	C700fa12be0b1c01e6a32044dd2813351.png showed the login page of Pinterest along with the user id along with few other thumbnails
	2	Internet Explorer browsing history	Firefox_Image.001/Partition 1/NONAME [NTFS]/[root]/Users/FTKuser/AppData/Local/Microsoft/Windows/WebCache/WebCacheV01.dat	Shows the images downloaded and uploaded from and to Pinterest. The visit count and Visit date corresponds to the time the post was uploaded or downloaded in the logs of evidence creation process
	3	Message	Firefox_Image.001/Partition 1/NONAME [NTFS]/[root]/Users/FTKuser/AppData/Local/Mozilla/Firefox/Profiles/7dl333s7.default/cache2/entries/029694E1538C52F75B77EC6421D9BA499EA96EC1	Pinterest image message between users Alice well, Bumble bee and Adam bomb were recovered in the cache entries folder. 029694E1538C52F75B77EC6421D9BA499EA96EC1 shows an image of koala bear which was shared by Alice well to Bumble bee
	4	Post	Firefox_Image.001/Partition 1/NONAME [NTFS]/[root]/Users/FTKuser/AppData/Local/Mozilla/Firefox/Profiles/7dl333s7.default/cache2/entries/053A41D688D6C8B724ADD72985727A7AE1DB88F27	The image that was shared in the Pinterest board of Alice well named Niki was recovered

Table 12 Firefox—indexed search results for Tumblr

Tumblr	S. No	Artifacts	Location	Details
	1	Slack space	Firefox_Image.001/Partition 1/NONAME [NTFS]/[root]/ Users/FTKuser/AppData/Local/Mozilla/Firefox/ Profiles/7dl333s7.default/cache2/entries/ 279D49DC5C992B42071D77837CDE35C996FDF7E7.FileSlack	Message between Tumblr users' yourtechnicallycrispystudent and ultrasweetbouquetcrusade was recovered. It shows the link of the image and the recipient of the image is ultrasweetbouquetcrusade
	2	Slack space	Firefox_Image.001/Partition 1/NONAME [NTFS]/[root]/ Users/FTKuser/AppData/Local/Mozilla/Firefox/ Profiles/7dl333s7.default/cache2/entries/ 949CEE42F38DBAFB3222459A98961DFA95C0DD64.FileSlack	The file shows that an image message was sent to recipient ultrasweetbouquetcrusade
	3	Thumbnails	Firefox_Image.001/Partition 1/NONAME [NTFS]/[root]/ Users/FTKuser/AppData/Local/Mozilla/Firefox/ Profiles/7dl333s7.default/thumbnails/	B199f5a3c825e8fd27418ec522fd3e23.png of a gif post made on Tumblr was found along with few images from the blogs visited by the user
	4	Internet Explorer browsing history	Firefox_Image.001/Partition 1/NONAME [NTFS]/[root]/ Users/FTKuser/AppData/Local/Microsoft/ Windows/WebCache/WebCacheV01.dat	Shows the images downloaded and uploaded from and to Tumblr The visit count and Visit date corresponds to the time the post were uploaded or downloaded in the logs of evidence creation process
	5	Message	Firefox_Image.001/Partition 1/NONAME [NTFS]/[root]/ Users/FTKuser/AppData/Local/Mozilla/Firefox/ Profiles/7dl333s7.default/cache2/entries/ 03473A9DF7E05409BC7F67FD1C6C8477435C06AB	Tumblr image message between users' yourtechnicallycrispystudent and Captainnumberonebouquetlover was recovered

(continued)

Table 12 (continued)

6	Message	Firefox_Image.001/Partition 1/NONAME [NTFS]/[root]/Users/FTKuser/AppData/Local/Mozilla/Firefox/Profiles/7dl333s7.default/cache2/entries/0376B0645F3BF3A74095EBC44C659856500BB5FF	Tumblr image message between users' yourtechnicallycrispystudent and ultrasweetbouquetcrusade was recovered
7	post	Firefox_Image.001/Partition 1/NONAME [NTFS]/[root]/Users/FTKuser/AppData/Local/Mozilla/Firefox/Profiles/7dl333s7.default/cache2/entries/583FB6BDAAA44ACCF5775062FB6694BE19AFED60/Untitled0	Link of the image posted on Tumblr was recovered
8	Video posted	Firefox_Image.001/Partition 1/NONAME [NTFS]/[root]/Users/FTKuser/AppData/Local/Mozilla/Firefox/Profiles/7dl333s7.default/cache2/entries/71B0F5E3B256692F1E83B83FABCEF3A1F16F08A6	The video that was shared in Tumblr was recovered
9	GIF posted	Firefox_Image.001/Partition 1/NONAME [NTFS]/[root]/Users/FTKuser/AppData/Local/Mozilla/Firefox/Profiles/7dl333s7.default/cache2/entries/06A5E2C2FC12CB1223EDE45154ABC0ED00B72A36	The GIF image that was shared in Tumblr was recovered
10	Messages	Firefox_Image.001/Partition 1/NONAME [NTFS]/[root]/Users/FTKuser/AppData/Local/Mozilla/Firefox/Profiles/7dl333s7.default/cache2/entries/1037C8B889AE4DF63F5EAE009ADF3A926AE3F4CA	Three message headers showing the participants of a chat were recovered
		Firefox_Image.001/Partition 1/NONAME [NTFS]/[root]/Users/FTKuser/AppData/Local/Mozilla/Firefox/Profiles/7dl333s7.default/cache2/entries/1FE8C034E53FA2A4E9C7859C005D3D650EEBF65B	
		Firefox_Image.001/Partition 1/NONAME [NTFS]/[root]/Users/FTKuser/AppData/Local/Mozilla/Firefox/Profiles/7dl333s7.default/cache2/entries/95716899965ED4AFBB9C9DCCCE0253805E8DE0AA4	

Table 13 Chrome—indexed search results

Pinterest	S. No	Artifacts	Location	Details
	1	Log-in credentials	[NTFS]/[root]/Users/ FTKuser/AppData/Local/Google/ Chrome/User Data/Default/Login Data	Email addresses (usernames) and Passwords (encrypted)
Tumblr	1	Log-in credentials	[NTFS]/[root]/Users/ FTKuser/AppData/Local/Google/ Chrome/User Data/Default/Login Data	Email addresses (usernames) and Passwords (encrypted)

with data from the physical memory. This goes to show that not only the versions of the browsers used, but also the type of operating system utilized by a user are important to an OSN investigation process. Also interesting, was the irregularity of downloaded images that were made using Firefox being found in the IE history, and some browsing history appearing in the system files.

There is no doubt that the artifacts found would prove very useful in legal cases as discussed above or whenever it is needed. Also, the application of the case's context to these artifacts would possibly lead to further findings. The consistency of these artifacts, i.e. would they always be found in these locations we have found them in cannot be ascertained, but they would provide a lead. A thorough and comprehensive investigation of the acquired evidence through searching and checking the well-known web artifacts location should not be underestimated.

10 Conclusion

This paper has been able to expose some of the challenges faced by the investigation of OSNs and current methodologies applied today in the investigation process. The use of the cutting-edge tools used today has also been demonstrated, and numerous artifacts have been found. Nonetheless, there is still much work to be done.

We are not far from the truth when we say microblogging and OSNs have formed a large part in our lives all together, and so, make major contribution to life's events, and should not be ignored.

Appendix

A. **Firefox**

See Fig. 2.

Results

Type	Data	Description	Found In
Password	elamathi	n582419@mvrht.net, WEBSITEhttps://www.pinterest.com	---
Password	elamathi@2017	n579359@mvrht.net, WEBSITEhttps://www.tumblr.com	---
Password	elamathi@2017	n581640@mvrht.net, WEBSITEhttps://www.tumblr.com	---

Fig. 2 Decrypted user id and password using PRTK

Fig. 3 c700fa12be0b1c01e6a3204dd2813351

Internet Explorer Browse History

Visit Date	Visit Count	
10/17/2017 6:22:36 PM -0500	2	file:///C:/Users/FTKuser/Downloads/cutedog.jpg
10/20/2017 5:15:06 PM -0500	3	file:///C:/Users/FTKuser/Downloads/ChickenTikka.gif
10/20/2017 5:38:54 PM -0500	2	file:///C:/Users/FTKuser/Downloads/baby.jpg
10/17/2017 5:39:49 PM -0500	1	file:///C:/Users/FTKuser/Downloads/Leopard.jpg
9/27/2017 4:57:09 PM -0500	1	file:///C:/Users/FTKuser/Downloads/MagnumEffectGif.gif
10/20/2017 5:02:49 PM -0500	1	file:///C:/Users/FTKuser/Downloads/Cute.jpg
9/27/2017 4:50:39 PM -0500	1	file:///C:/Users/FTKuser/Downloads/PuppyNgirl.jpg
10/2/2017 6:54:33 PM -0500	1	file:///C:/Users/FTKuser/Downloads/uinacup.jpg
9/27/2017 5:13:47 PM -0500	2	file:///C:/Users/FTKuser/Downloads/Roses.jpg
10/20/2017 5:14:19 PM -0500	1	file:///C:/Users/FTKuser/Downloads/amazon.jpg
10/20/2017 4:48:30 PM -0500	2	file:///C:/Users/FTKuser/Downloads/YummyFries.jpg

Fig. 4 Internet explorer browsing history

Pinterest

1. Thumbnail image:
   Firefox_Image.001/Partition   1/NONAME   [NTFS]/[root]/Users/FTKuser
   /AppData/Local/Mozilla/Firefox/Profiles/7dl333s7.default/thumbnails/ (Fig. 3)
2. Internet Explorer browsing history
   Firefox_Image.001/Partition   1/NONAME   [NTFS]/[root]/Users/FTKuser
   /AppData/Local/Microsoft/Windows/WebCache/WebCacheV01.dat (Fig. 4)
3. Message

Fig. 5 Message viewed in Pinterest window

Fig. 6 The dashboard and image viewed in Pinterest

Firefox_Image.001/Partition   1/NONAME   [NTFS]/[root]/Users/FTKuser
/AppData/Local/Mozilla/Firefox/Profiles/7dl333s7.default/cache2/entries
/029694E1538C52F75B77EC6421D9BA499EA96EC1 (Fig. 5)
4. Post
    Firefox_Image.001/Partition   1/NONAME   [NTFS]/[root]/Users/FTKuser
    /AppData/Local/Mozilla/Firefox/Profiles/7dl333s7.default/cache2/entries
    /053A41D688D6C8B724ADD72985727A7AE1DB8F27 (Fig. 6)

Tumblr

1. Message in slack space
    Firefox_Image.001/Partition   1/NONAME   [NTFS]/[root]/Users/FTKuser
    /AppData/Local/Mozilla/Firefox/Profiles/7dl333s7.default/cache2/entries
    /279D49DC5C992B42071D77837CDE35C996FDF7E7.FileSlack   (Figs.   7
    and 8)
2. Message in slack space

"key":"ultrasweetbouquetcrusade.tumblr.com","can_pixelate_avatar":false,"avatar_url":"https://assets.tumblr.com/images/default_avatar/octahedron_open_6 4.png","is_active":true,"tags":[]}],"messages":{"data":[{"type":"IMAGE","ts":"1505946682928", "participant":"ultrasweetbouquetcrusade.tumblr.com","images" :[{"alt_sizes":[{"url":"https://68.media.tumblr.com/52aeb92be74e3224f903fdac43c5677e/tumblr_messaging_owlnw99OJI1v8rppg_1280.jpg","width":1024,"h eight":768},{"url":"https://68.media.tumblr.com/52aeb92be74e322

Fig. 7 Recovered message

Fig. 8 The link represents the above message image

Cache-Control: no-cache, must-revalidate
Content-Type: image/gif
Expires: Sat, 26 Jul 1997 05:00:00 GMT
Content-Length: 23
original-response-headersCache-Control: no-cache, must-revalidate
Content-Type: image/gif
Expires: Sat, 26 Jul 1997 05:00:00 GMT
Content-Length: 23
Connection: keep-alive
net-response-time-onstart414net-response-time-
onstop414{ fit!! ⬜ I°E,ᴸ   YI$#YÒÏ$>>"LYI. ⁺]:https://safe.txmblr.com/svc/embed/iframe/ultrasweetbouquetcrusade/165745933417?w=540&h=304necko: classified1strongly-framed1security-
infoFnhlIAKWRHGAIo+ESXykKAAAAAAAAAAwAAAAAAAAEaphJoJH6pBabD5gSnsfLHeAAQAAgAAAAAAAAAAAAAAAAAAAB4+vFIJp5wRkeyPxAQ9RJGKPqbqVvKO0mKuJ8ec8o/uhmCJImkVxP+7sgiYWmM
t8FvcOXmlQiTNWFiWIrbpbqgwAAAAAAAAgEMIIIADCCBuigAwIBAgIQAy9R4SkmcPsWqWmmK8J5VDANBgkqhkiG9w0BAQsFAD8wMQswCQYDVQQGEwJVUzEVMBMGA1UEChMMRGInaUNIcnQgSW5jJMRkwF
wYDVQQLExB3d3cuZGInaWNIcnQuY29tMS8wLQYDVQQDEyZEaWdpQ2VydCBTSEEyIEhpZ2ggQXNzdXJhbmNlIFNlcnZlciBDQTAeFw0xNjEyMTkwMDAwMDBaFw0xODAxMDMxMjAwMDBaMGAxCzAJBgNVBAY
TAlVTMREwDwYDVQQIEwhOZXcgJW9ZyazERMA8GA1UEBxMITmV3IFivcmsxFDASBgNVBAoTC1R1bWJsciBJbmMuMRUwEwYDVQQDDAwqLnR4bWJsci5jb20wwggEiMA0GCSqGSIb3DQEBAQUAA4IBDwAwgg
EKAoIBAQDe9BNRagkX85Yoxj3js80rsJG77oO7OwiMxVjsv0OmPtEtjCM4s6ZibrwzyGXFs4+WDtbOhCwWT8rzJ7dxXyLW53KMeLsHloGiftGAxoVTCz7DziNs4DlEPHM7680nHS1f1ZVs0GRCWdWwCJI0ntWkYkAfO
O9njJbp8sIRsHtbvUwDPhJ/r6u6pRJK+BWz3YZGqGSlHGcaPEwhsuB04Rox4QIR4XTH

Fig. 9 Recovered message showing recipient and type of message

Firefox_Image.001/Partition   1/NONAME   [NTFS]/[root]/Users/FTKuser /AppData/Local/Mozilla/Firefox/Profiles/7dl333s7.default/cache2/entries /949CEE42F38DBAFB3222459A98961DFA95C0DD64.FileSlack (Fig. 9)
3. Post found in Thumbnail
Firefox_Image.001/Partition   1/NONAME   [NTFS]/[root]/Users/FTKuser /AppData/Local/Mozilla/Firefox/Profiles/7dl333s7.default/thumbnails/ (Fig. 10)
4. Internet Explorer browsing history
Firefox_Image.001/Partition   1/NONAME   [NTFS]/[root]/Users/FTKuser /AppData/Local/Microsoft/Windows/WebCache/WebCacheV01.dat
5. Message
Firefox_Image.001/Partition   1/NONAME   [NTFS]/[root]/Users/FTKuser /AppData/Local/Mozilla/Firefox/Profiles/7dl333s7.default/cache2/entries

Fig. 10 Thumbnail image of a GIF post made by the user

Fig. 11 Gif received viewed in Tumblr

/03473A9DF7E05409BC7F67FD1C6C8477435C06AB (Fig. 11)

6. Message
   Firefox_Image.001/Partition   1/NONAME   [NTFS]/[root]/Users/FTKuser
   /AppData/Local/Mozilla/Firefox/Profiles/7dl333s7.default/cache2/entries
   /0376B0645F3BF3A74095EBC44C659856500BB5FF (Fig. 12)

7. Post
   Firefox_Image.001/Partition   1/NONAME   [NTFS]/[root]/Users/FTKuser
   /AppData/Local/Mozilla/Firefox/Profiles/7dl333s7.default/cache2/entries
   /583FB6BDAAA44ACCF5775062FB6694BE19AFED60/Untitled0 (Fig. 13)

8. Video posted

Fig. 12 Image sent viewed in Tumblr

ro is_photo post_tumblelog_ccda6b921918a9ca39eb104e1223/a79e not_mine is_original with_permalink no_source generic_source post--owner_flagged_nsfw reblog_ui_refresh U+0022 id= U+0022p
data-serve-id= U+0027ab988bdf2eaf0efc29f2a07fabc45934 U+0027 data-json= U+0027{"id":"166549513367","type":"photo","root_id":&q
xt;,"name":"cutefunnybabyanimals","cname":"","description":"Funny and cute animals around the world","description_sani
nnycuteanimals","likes":true,"share_following":false,"is_blogless_advertiser":false,"is_private":false,"is_group":false,"customizat
tefunnybabyanimals","embed_key":"eUfzGvSZhJfb-qr53S_eUg","embed_did":"c2c4335b8fa1e4702ef3e6e9fe7f01ff849e06508","post_id&quot
:is_private":0,"has_user":false,"has_facebook":false,"twitter_username":"","permalink_label":"Permalink","show_re
tps%3A%2F%2F78.media.tumblr.com%2F393a6b9b265efa2a588f5cedd4dc5666%2Ftumblr_oy1lroTAva1tzkxwxo1_500.jpg","name":"pinterest-share-dialog-166549513367&q
\u0022,\u0022root_id\u0022:\u0022166549513367\u0022,\u0022tumblelog\u0022:\u0022cutefunnybabyanimals\u0022,\u0022tumblelog-key\u0022:\u0022oSkLAem8P\u0022,\u0022tumblelog-da
i\u0022:\u0022Funny and cute animals around the world\u0022,\u0022title\u0022:\u0022Funnycuteanimals\u0022,\u0022likes\u0022:true,\u0022share_following\u0022:false,\u0022is_blogless_ad
data\u0022:false,\u0022reblog_key\u0022:\u0022qsVdCbW2\u0022,\u0022is_reblog\u0022:false,\u0022is_mine\u0022:false,\u0022liked\u0022:false,\u0022sponsored\u0022:\u0022,\u002
,549513367\u0022,\u0022post_url\u0022:\u0022https:\\\\\/\cutefunnybabyanimals.tumblr.com\\\/post\\\/166549513367\\\/6-weeks-vs-6-months-via-raww\u0022,\u0022post_tiny_url\u0022:\u0
,\/button\\\/?url=https:\\\/\\\/cutefunnybabyanimals.tumblr.com\\\/post\\\/166549513367\\\/6-weeks-vs-6-months-via-raww&description=&media=https%3A%2F%2F78.media.tumblr.co
sls\u0022,\u0022reblog-key\u0022:\u0022qsVdCbW2\u0022,\u0022direct-video\u0022:\u0022\u0022,\u0022is-animated\u0022:false,\u0022serve-id\u0022:\u0022ab988bdf2eaf0efc29f2a07fabc45
U+0022 U+003E
_inner U+0022 U+003E

il_container U+0022 style= U+0022background-image: url(U+0027https://78.media.tumblr.com/393a6b9b265efa2a588f5cedd4dc5666/tumblr_oy1lroTAva1tzkxwxo1_250sq.jpg U+0027); U+0022

Fig. 13 Image link of the post recovered

Firefox_Image.001/Partition   1/NONAME   [NTFS]/[root]/Users/FTKuser
/AppData/Local/Mozilla/Firefox/Profiles/7dl333s7.default/cache2/entries
/71B0F5E3B256692F1E83B83FABCEF3A1F16F08A6 (Fig. 14)
9. Gif posted
Firefox_Image.001/Partition   1/NONAME   [NTFS]/[root]/Users/FTKuser
/AppData/Local/Mozilla/Firefox/Profiles/7dl333s7.default/cache2/entries
/06A5E2C2FC12CB1223EDE45154ABC0ED00B72A36 (Fig. 15)
10. Messages showing participants
Firefox_Image.001/Partition   1/NONAME[NTFS]/[root]/Users/FTKuser
/AppData/Local/Mozilla/Firefox/Profiles/7dl333s7.default/cache2/entries/
Figures 16 and 17 shows the participants of a chat and the time they were created
matched the time in the evidence creation process logs.

Fig. 14 Video posted viewed in Tumblr

Fig. 15 Gif posted viewed in Tumblr

"tT~T*} k mAZ²}-xfF¹◆,,8!ë q|,>PJ 8F] H⫶ Yv CL}}%J6BsDmY ûû@,~
:https://www.tumblr.com/svc/conversations/messages?participants%5B%5D=yourtechnicallycrispystudent.tumblr.com&participants%5B%5D=captainnumberonebouquetlover.tumblr.com&participant=
yourtechnicallycrispystudent.tumblr.com&_=1508357340448 necko:classified strongly-framed security-info FnhllAKWRHGAlo+ESXykKAAAA AAAAwAAAA AAAAEaphjojH6pBabDSgSnsfLHeAAQAA
bDSgSnsfLHeAQAAgAAAA AAAAB4vFIJp5wRkeyPxAQ9RJGKPqbqVv yPxAQ9RJGKPqbqVvKO0mKuⅡ8ec8o/uhmCJImkVxP+7sgiYWmMt8FvcOXmlQiTNWFiWIrbpbqg lQiTNWFiWIrbpbqgwAAAA
AAAAe6MⅢHtjCCBp6gAwIBAgIQC+vDdvDERz93xvT/ul+G6zANBgkqhkiG9w0BAQsFADBwMQswCQY BAQsFADBwMQswCQYDVQQGEwJVUzEVMBM GA1UEChMMRGlnaUN lcnQgSW5jMRkwFwY
DVQQLExB3d3cuZGl naWNlcnQuY29tMS8 wLQYDVQQDEyZEaWd pQ2VydCBTSEEyIEh pZ2ggQXNzdXJhbmN lIFNlcnZlciBDQTA eFw0xNzA5MTEwMDA wMDBaFw0xNzEwMJY xMjAwMDBaMGYxCzA
J8gNVBAYTAIVTMQs wCQYDVQQIEwJDQTE SMBAGA1UEBxMJU3V ubnl2YWxlMRQwEgY DVQQKDAtZYWhvbyE gSW5jLjEgMB4GA1U EAvwxXKi5nbG9iYWw tcG9wLnR1bWJsci5 jb20wggEiMA0GCSq
GSIb3DQEBAQUAA4I BDwAwggEKAoIBAQD SsJyVmPVMhmrnQ9cb7 7b7g96QiuA0oOlZX MBnG7fJK+r3AGmG0LULm+yyT+Z85gc8+E2V65AkVq6t5g1dM7d5ApwvFzS1okR6v
7d5ApwvFzS1okR6wbEa4wDIb3sNKk8I+tQyxc/wBpHW43Y3v6KQneXEJnHZBDP2GHRtf3 EJnHZBDP2GHRtf3bHOsMdKO7H4VlG+FskiЖd7Zm1fNT3TIshADVsdSufSG4tn
IshADVsdSufSG4tnwunJ048+1keTnpKykm1ip6YSClue7xLVgXFygsF8 Clue7xLVgXFygsF8AFeEQrW+PuqfEgcXluycZyslgwrkERTdXvE2oZor gwrkERTdXvE2oZoroyqGpXKtgGPJaj2f xkUCoKklkVjSlobA
SNzHSAi4cⅡaZJlG 8E1uqYuJ11/pH/AgMBAAGJggRUMⅢEUDAfBgNVHSMEGDAW UDAfBgNVHSMEGDAWggBRRaP+QrwJHdTzM2WVkYqISuFlyOzAd8gNVHQ4 SuFlyOzAd8gNVHQ4EFgQUGHd9g6ujHTH
h6lkfgJkU8tc8fMA wgYsGA1UdEQSBgzC BglDXi5nbG9iYWw tcG9wLnR1bWJsci5 jb22CCnR1bWJsci5 jb22CDnd3dy50dW1 ibH1uY29tggShcGk udHVtYYmxyLmNvbYI Ud3d3LWh0dHAyLnR 1bWJsci5jb22CFGF

Fig. 16 Shows message participants

@8P{VU kA Q+! 4`!8S2 <~%=S: imXga}u\ &B:h<$!xY(Lⅱ j :XbR S ⫶ @P)
:https://www.tumblr.com/svc/conversations/messages?participants%5B%5D=yourtechnicallycrispystudent.tumblr.com&participants%5B%5D=ultrasweetbouc
uetcrusade.tumblr.com&participant=yourtechnicallycrispystudent.tumblr.com&_=1508534792029 necko:classified strongly-framed security-info
FnhllAKWRHGAlo+ESXykKAAAA AAAAwAAAA AAAAEaphjojH6pBabDSgSnsfLHeAAQAA bDSgSnsfLHeAAQAAgAAAA AAAAB4vFIJp5wRkeyPxAQ9RJGKPqbqVv
yPxAQ9RJGKPqbqVvKO0mKuⅡ8ec8o/uhmCJImkVxP+7sgiYWmMt8FvcOXmlQiTNWFiWIrbpbqg lQiTNWFiWIrbpbqgwAAAA
AAAAe6MⅢHtjCCBp6gAwIBAgIQC+vDdvDERz93xvT/ul+G6zANBgkqhkiG9w0BAQsFADBwMQswCQY BAQsFADBwMQswCQYDVQQGEwJVUzEVMBM
GA1UEChMMRGlnaUN lcnQgSW5jMRkwFwY DVQQLExB3d3cuZGl naWNlcnQuY29tMS8 wLQYDVQQDEyZEaWd pQ2VydCBTSEEyIEh pZ2ggQXNzdXJhbmN
lIFNlcnZlciBDQTA eFw0xNzA5MTEwMDA wMDBaFw0xNzEwMJY xMjAwMDBaMGYxCzA J8gNVBAYTAIVTMQs wCQYDVQQIEwJDQTE SMBAGA1UEBxMJU3V
ubnl2YWxlMRQwEgY DVQQKDAtZYWhvbyE gSW5jLjEgMB4GA1U EAwwxXKi5nbG9iYWw tcG9wLnR1bWJsci5 jb20wggEiMA0GCSq GSIb3DQEBAQUAA4I

Fig. 17 Shows message participants

B. Screenshots of IE v11 Artifacts found as shown in AccessData FTK

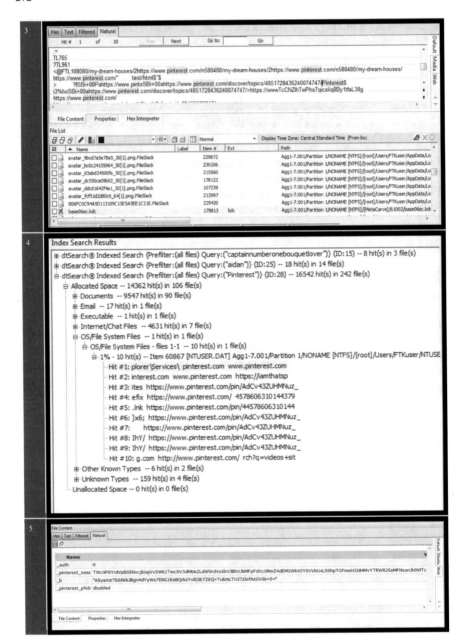

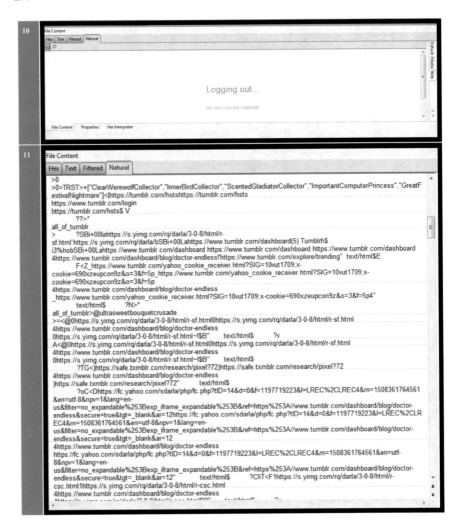

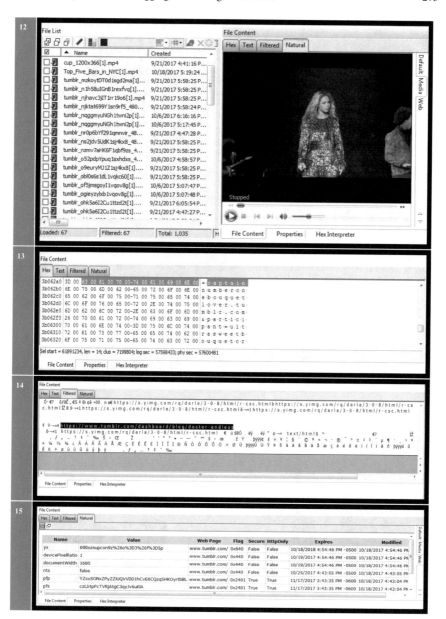

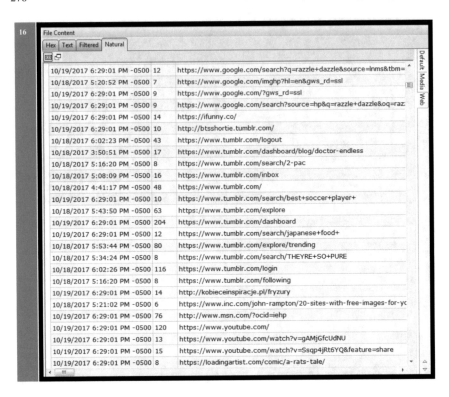

C. Google Chrome Analysis Screenshots

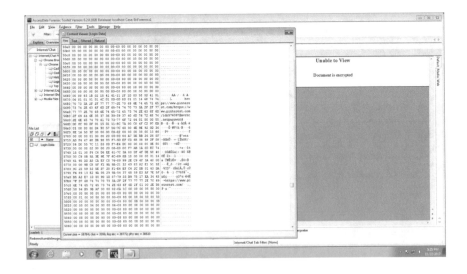

1. Pinterest Log-in Credentials

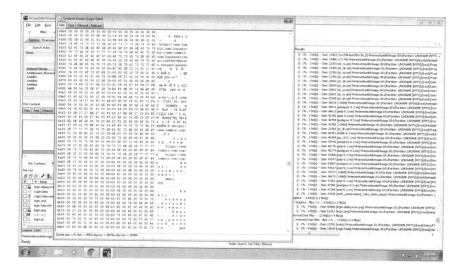

2. Tumblr Log-in Credentials

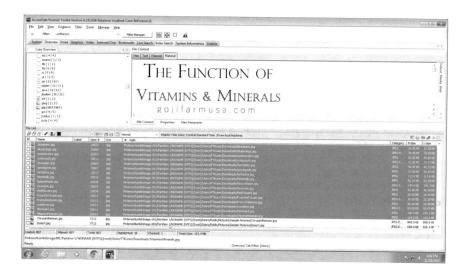

3. Contents of Downloads Folder

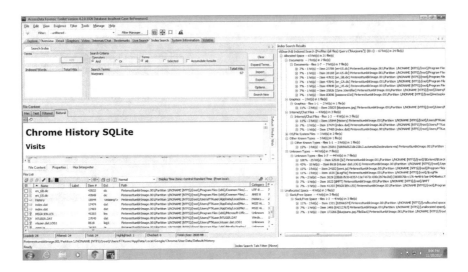

4. Location of Google Chrome History Folder

References

1. Dardick, G.S., La Roche, C.R., Flanigan, M.A.: Blogs: anti-forensics and counter anti-forensics (2007)
2. Zainudin, N.M., Merabti, M., Llewellyn-Jones, D.: Online social networks as supporting evidence: A digital forensic investigation model and its application design. In: 2011 International Conference on Research and Innovation in Information Systems, pp. 1–6 (2011)
3. Taylor, M., et al.: Forensic investigation of social networking applications. Netw. Secur. **11**(2014), 9–16 (2014)
4. Zainudin, N.M., Merabti, M., Llewellyn-Jones D.: A digital forensic investigation model for online social networking. In: Proceedings of the 11th Annual Conference on the Convergence of Telecommunications, Networking and Broadcasting, Liverpool (2010)
5. Mulazzani, M., Huber, M., Weippl, E.: Social network forensics: tapping the data pool of social networks. https://www.semanticscholar.org/paper/Social-Network-Forensics-Tapping-the-Data-Pool-of-Mulazzani-Huber/3bc38655af392e1489eeeeab9a7a2fa8a45c5edb. Accessed 13 Sep 2017
6. Oh, J., Lee, S., Lee, S.: Advanced evidence collection and analysis of web browser activity. Digit. Investig. **8**(Supplement), S62–S70 (2011). ISSN 1742-2876, https://doi.org/10.1016/j.diin.2011.05.008, http://www.sciencedirect.com/science/article/pii/S1742287611000326
7. Pereira, M.T.: Forensic analysis of the Firefox 3 Internet history and recovery of deleted SQLite records. Digit. Investig. 5(3–4), 93–103 (2009). ISSN 1742-2876, https://doi.org/10.1016/j.diin.2009.01.003, http://www.sciencedirect.com/science/article/pii/S1742287609000048
8. Rahman, S., Khan, M.N.A.: Digital forensics through application behavior analysis. Int. J. Mod. Educ. Comput. Sci. **8**(6), 50 (2016)
9. https://support.mozilla.org/en-US/kb/profiles-where-firefox-stores-user-data#w_what-information-is-stored-in-my-profile

10. Seigfried-Sellar, K., Leshney, S.: The intersection between social media, crime, and digital forensics: #WhoDunIt?. ResearchGate. https://www.researchgate.net/publication/301266641_The_intersection_between_social_media_crime_and_digital_forensics_WhoDunIt. Accessed 13 Sep 2017

11. Rathod, D.M.: Web browser forensics: Google Chrome. Int. J. Adv. Res. Comput. Sci. **8**(7). https://login.libweb.lib.utsa.edu/login?url=https://search-proquest-com.libweb.lib.utsa.edu/docview/1931129713?accountid=7122 (2017)

12. Jones, K.J.: Forensic analysis of internet explorer activity files. Forensic Analysis of Microsoft Windows Recycle Bin Records (2003)

13. Ohana, D.J., Shashidhar, N.: Do private and portable web browsers leave incriminating evidence?: A forensic analysis of residual artifacts from private and portable web browsing sessions. EURASIP J. Inf. Secur. **2013**, 1–13 (2013). http://dx.doi.org.libweb.lib.utsa.edu/10.1186/1687-417X-2013-6

14. Akbal, E., Günes, F., Akbal, A.: Digital forensic analyses of web browser records. JSW **11**(7), 631–637 (2016)

15. Said, H., et al.: Forensic analysis of private browsing artifacts. In: International conference on Innovations in information technology (IIT). IEEE (2011)

16. Jang, Y.-J., Kwak, J.: Digital forensics investigation methodology applicable for social network services. Multimed. Tools Appl. 74(14), 5029–5040 (2015)

17. Yang, T.Y., et al.: Windows instant messaging app forensics: Facebook and Skype as case studies. PloS One **11**(3), e0150300 (2016)

Social Media Data in Digital Forensics Investigations

Ashleigh Powell and Cydnee Haynes

Abstract Evidence collected from social media presents valuable information that should not be overlooked. Evidence can be captured from social media using multiple methods including searching publicly viewable content, reviewing content metadata, soliciting and investigating interactions with other users, and utilizing legal holds. After the evidence is gathered, it can be utilized in various ways. Social media evidence can be used to create a timeline of events, show intent or conspiracy, and establish connections between persons. Digital forensics investigations can be used to collect such evidence. The evidence accumulated from social media is far-reaching and widely advantageous. However, there are many legal issues that can affect the collection and ultimate legal admissibility of this evidence. Evidence must be collected using careful, correct procedures and in a manner that ensures its integrity. The ethical implications of the collection of social media evidence also plays a role in these digital forensics investigations. These issues can present adverse circumstances for cases. Nonetheless, social media offers various avenues for the collection and use of its data as evidence within a digital forensics investigation. The advantageous and disadvantageous aspects of this use are discussed in this chapter.

1 Introduction

The term social media is used to describe the numerous social networking websites and applications that are used to create, distribute, and receive content. These tools have affected the ways that individuals share and experience life. Individuals may use social media to gain knowledge of new trends, give input on disparate topics, or introduce new products and services to the global market. People tend to use social media as a means of staying connected to the world by sharing their day-to-day experiences. Individuals will use social media to connect with the global environment and enhance their reach and influence, all within an affordable price range. This inex-

A. Powell (✉) · C. Haynes
Department of Information System and Cyber Security,
University of Texas at San Antonio, San Antonio, TX, USA
e-mail: ashleigh.powell@hotmail.com

© Springer Nature Switzerland AG 2020
X. Zhang and K.-K. R. Choo (eds.), *Digital Forensic Education*,
Studies in Big Data 61, https://doi.org/10.1007/978-3-030-23547-5_14

pensive yet expansive capability has proliferated the use of social media in a variety of ways. Besides individual consumer use, social media websites and applications can be used to conduct business research, engage in e-commerce, communicate with business partners, and recruit new employees. Social media acts as an advertisement channel for many businesses in many different industries. Companies use these tools to engage with current and potential customers. Companies can interact with these consumers on a more personal level which can make their messages more effective. Individuals and businesses alike engage in social media use in a myriad amount of ways.

Social media use has significantly increased in recent years. This continuous growth exemplifies the need for a better understanding of the implications of these websites and applications in digital forensics investigations. "A digital forensics investigation is a process that uses science and technology to analyze digital objects and that develops and tests theories, which can be entered into a court of law, to answer questions about events that occurred" [5, p. 13]. A digital object can be anything from a name or an image down to the computational methods or properties for that name or image used within a computer. Digital objects quickly become digital evidence when gathered during digital forensics investigation for use in a court of law. Digital evidence obtained from social media can be used to the advantage of law enforcement in a variety of ways. Social media generates enormous amounts of data every day. This creates an almost endless amount of informative and beneficial data. This data can be recovered and analyzed for use within a digital forensics investigation. Since the use of social media is constantly increasing, data is being generated exponentially. This expansion may complicate its use in digital forensics investigations. These investigations must now consider the various ways information is added and disseminated through social media. The artifacts found and collected from social media have had controversial settings in a court of law. However, some evidence has been used favorably to solve a case or create another lead.

Although social media is largely used to conduct friendly activities, it can also be used to facilitate malicious tasks. Criminal behavior and evidence of crimes can be found throughout social media. The data found on these websites and applications can imply intent when used in a criminal case. In addition, social media can be used to share criminal activity. There have been many cases where the after-effects of a crime or a crime in progress have been broadcasted through social media. This data can also be used as evidence. Individuals may access social media websites and applications using various devices, but it is common for access to be gained via smartphones. Therefore, these devices must be investigated properly to ensure careful preservation and collection of evidence. Many crimes can be somehow connected to social media in today's digital age. The ethical and legal admissibility of this evidence is often debated, such as when the data in retrieved from public social media posts. Law enforcement will have to fathom these implications in digital forensics investigations.

2 Related Work

Social media has become one of the primary vectors through which information is shared in today's digital environment. These sites and applications allow users to express themselves and their personalities [8]. According to Parveen et al. [23], many organizations will use social media to generate revenue and impact users. However, we live in a society that uses social media in many different ways and not all of them are for good [27]. Social media can be used to conduct malicious activities as explained by Al-khateeb et al. [1]. Today, almost anyone and everyone can peer into a user's "private" life simply by searching for them on social media [29]. Al-khateeb et al. [1] asserted that social media applications and websites can create networks for deviant use. They termed these networks as deviant hacker networks (DHNs) and described them to be the communication of malevolent users over online social media networks. The social networking service examined in this study was Twitter. They discovered the commonly used websites, domains, accounts, and hashtags of these DHNs. One group identified was the infamous hacker group called Anonymous. They identified how often these deviant groups communicated with each other and how they may have worked together through the use of social media. Much of this information was found using publicly available content. Bradbury [3] explains that harvesting such data is beneficial to many investigations in several ways.

Another example of the negative use of social media is within e-commerce and white collar crimes. Digital forensics can be used to help these types of cases. Fianyi [7] asserted that these financially-motivated crimes are increasing as the inevitable function of conducting business online widens. He investigated the risks, costs, and damages these businesses and individuals can face by utilizing the Internet in this way. Kwahk and Ge [20] explain that social media interactions positively affect social influence which in turn affects customer visit and purchase intentions within e-commerce. These influences and interactions could be traced to find connections and motives. Social media has also been used to facilitate other cyber-crimes ranging from embezzlement to fraud [18]. Karabiyik et al. [18] provided a review on how to better understand crime and terror network structures. These structures are the foundations of where crimes using social media begin. They asserted that one must first understand the ranks of individuals, relationships, and roles. Understanding these aspects can help prevent possible attacks. They also went on to explain that attacks using social media and its networks are critical because often ordinary users are not fully aware of who or what they may be interacting with online. This provides a gateway for identity theft and other attacks, including terrorism. Karabiyik et al. [18] also explained that these crimes are always evolving, and that research on social networks and digital forensics must be understood in order to prevent these criminal activities in their changing forms.

Online terrorism is a prevalent problem in today's society. According to Dean et al. [6], social media plays a big role in online terrorism. This research was focused on three of the most popular social media platforms used today. These are Facebook, Twitter, and YouTube. Their focal point is on "homegrown terrorism". This term can best be defined as committing terrorist acts in the offenders' home country against fellow citizens. Social media is often used to recruit, communicate, and train prospective members of terrorist organizations. Huey [16] explains that this is done through a practice known as political jamming. She defines political jamming as "the subversion of popular memes to propagate pro-terrorist messages" [16]. Kim [19] suggests that online political messaging influences a social media user's exposure to cross-cutting political points of view. These messages can be used to communicate radical views and beliefs, thereby further influencing a potential recruit. This is possible because social media provides a platform for online users to create communities, share information, and generate content to be used in virtually any way necessary to achieve their goals and desires, may they be good or bad.

Smartphones are a typical way in which social media is accessed, so these devices must be explored as offered by Al Mutawa et al. [2] and Walnycky et al. [30]. Al Mutawa et al. [2] referred to these devices as the "goldmine for forensic investigations". Research was conducted to determine if activities performed on social networking applications through smartphones were stored in the device's internal memory. The results detailed the amount, significance, and location of interesting data that was found on the tested iPhones and Android phones. It is important to note that the BlackBerry phones that were tested contained no traces of social networking activity. However, this work is still significant because iPhones and Android phones currently dominate the smartphone market. Data from social media applications on smartphones tend to be unencrypted when stored and transmitted. Walnycky et al. [30] discuss the advantages and disadvantages of this. It is said that this type of unsecured storage and transmission implies benefits for digital forensic examiners but creates issues of privacy for social media users.

Law enforcement agencies must learn to utilize information that can be found in or derived from social media as suggested by Ramwell et al. [24] and Seigfried-Spellar and Leshnay [26]. Seigfried-Spellar and Leshnay [26] discuss how social media, crime, and digital forensics go hand in hand. There are various sources that digital forensics can be conducted on. They list devices and tools that may aid in an investigation involving social media evidence. Trottier [28] explains that social media offers high visibility and easy access to its data. This benefits law enforcement by allowing greater efficiency in evidence collection. Ramwell et al. [24] provide techniques for officers to utilize while undertaking an investigation through an online presence. They provided points of reference that could be used in an investigation. Each point attributed to an identifiable context to be searched such as location, person, or ego. Use cases were provided to further solidify the benefits of using information provided by social media in an investigation. Cai et al. [4] propose methods on how to remove data from social media which can help law enforcement to understand how some criminals may try to hide evidence. However, evidence obtained from social media may not always be admissible in a court of law and should be collected carefully. Murphy and Fontecilla [22] discuss some of these issues. They discussed the proliferation of social media in recent years and provided examples of using social media in government investigations. Holt and San Pedro [13], Grimm et al. [10], as well as Hoffmeister [12] further discuss what makes social media evidence admissible and the barriers law enforcement may face when presenting such evidence in court. In addition, how social media evidence is obtained and used by law enforcement is often called into question or looked down upon as indicated by Hesketh and Williams [11].

The professional workplace can breed abuse of social media as well. Employees may use social media as a means to defame their employers or lie to get extra time off from work as described by Karabiyik et al. [18]. While some companies have policies in place to collect digital evidence that they have the legal right to, many do not have authority over evidence collected from external sources in which social media may fall under [15]. Jang and Kwak [17] and Huber et al. [14] seek to resolve this problem by proposing a method for conducting digital evidence collection on social media networks. Huber et al. [14] proposed using a web crawling application in combination with a third-party application to collect user data and metadata from these sites. The purchase of third-party applications for these types of investigations is another serious consideration. More sophisticated tools may need to be used as digital technology continues to advance. Jang and Kwak [17] describe the importance of collecting evidence in a secure manner to ensure its integrity. This is another important consideration as it can determine the admissibility of digital evidence in a court of law.

Digital evidence can be an integral part of any investigation. Social media evidence can be found in several forms including search engine results as explained by Xiang and Gretzel [31]. How that evidence is collected and preserved is very important and law enforcement may encounter various legal barriers. Marsoof [21] and Rubio [25] discuss the ethical and legal implications of privacy and laws such as the Electronic Communications Privacy Act. Law enforcement in most jurisdictions is likely to require a warrant before any evidence can be collected. A warrant aids in the persuasion of outside parties involved in an investigation to cooperate. A warrant is also useful in obtaining information from social media companies. Hubert [15] provides a case in which an individual may obtain information from a social media website versus how law enforcement may go about obtaining the same information. In the case, a representative from a private company requested data from Facebook, of which the social media network owner was not obligated to give. However, if it is discovered during an investigation that the suspect is in violation of the site's terms and conditions, the site owner runs the risk of incurring legal liability. In these incidents, more often than not, the owner will do what is right and cooperate with authorities. There are laws and regulations that law enforcement must abide by when it comes to utilizing social media evidence. Rubio [25] provides an explanation of the various laws and expectations that must be adhered to, but also considers the ethics that may get called into question. If procedures are followed correctly, the amount of digital evidence derived from social media that can be used in the court of law has no bounds. In summary, while social media can be a powerful tool in aiding a digital forensics investigation, it is not without its drawbacks. In this paper, we aim to provide an analysis of both these benefits and downsides.

For this paper, our analysis of the uses of social media evidence as well as its adversities included research from scholarly peer reviewed journals dating back to 2010 and across various databases through the University of Texas at San Antonio library. Attributes we looked for in selecting the articles for this paper included unbiased and supportable information on the crimes committed with social media, detail on how social media was used as evidence in the case, and regulations and procedures dictating how and when social media could be used as evidence. As a result of using digital forensics to investigate social media evidence being a relatively new field of interest, most articles highlighted its use in an ongoing criminal investigation. An additional common theme demonstrated amongst the articles contributing to our research was how social media can be used and abused by both criminals and the law (see Table 1).

Table 1 Related work publications

Author	Title	Key findings	Analysis
Al-khateeb, S., Conlan, K. J., Agarwal, N., Baggili, I., & Breitinger, F.	Exploring Deviant Hacker Networks (DNH) On Social Media Platforms	Social media presents a medium for hacker networks	Hacker networks can be discovered and analyzed for intelligence
Al Mutawa, N., Baggili, I., & Marrington, A	Forensic Analysis of Social Networking Applications on Mobile Devices	Data from social media applications can be found in the memory of smartphones	Social media data is recoverable from smartphones and can be used as evidence
Bradbury, D.	In Plain View: Open Source Intelligence	Open source intelligence offers a surprising amount of data	Open source intelligence aids investigations
Cai, Z., He, Z., Guan, X., & Li, Y.	Collective Data-Sanitization for Preventing Sensitive Information Inference Attacks in Social Networks	Proposed data sanitization methods for user profile and friend relations	There are ways of removing data from social media which can be used covertly
Dean, G., Bell, P., & Newman, J.	The Dark Side of Social Media: Review of Online Terrorism	Social media provides a stable platform for online terrorism, and various social media networks are used for this global crime	Social media has many functions in the realm of online terrorism
Fianyi, I. D.	Curbing Cyber-Crime and Enhancing E-Commerce Security with Digital Forensics	E-commerce presents significant risks for businesses	Businesses can use digital forensics to aid in financial crime investigations and deter cybercrime

(continued)

Table 1 (continued)

Author	Title	Key findings	Analysis
Gosling, S., Gaddis, S., & Vazire, S.	Personality Impressions Based on Facebook Profiles	Online social networks can communicate personality	This information can be utilized in an investigation
Grimm, P., Bergstrom L., & O'Toole-Loureiro, M.	Authentication of Social Media Evidence	Social media evidence authentication has become a very important process in litigation today	In order to be used in court, social media evidence must be proven authentic, amongst other requirements, for it to be admissible
Hesketh, I. & Williams, E.	A New Canteen Culture: The Potential to Use Social Media as Evidence in Policing	Police may use social media in a multitude of ways in an investigation	How police use social media in investigations draws attention from those who oppose the use and those who are for it
Hoffmeister, T.	The Challenges of Preventing and Prosecuting Social Media Crimes	There are many crimes that occur through the use of social media	Individuals who commit a crime with the use of social media may not always be punished
Holt, M. & San Pedro, V.	Social Media Evidence: What You Can't Use Won't Help You	Finding, preserving, and presenting social media evidence can be difficult	There are many considerations that must be taken into account when evidence is gathered from the Internet
Huber, M., Mulazzani, M., Leithner, M., Schrittwieser, S., Wondracek, G., & Weippl, E.	Social Snapshots: Digital Forensics for Online Social Networks	Proposed web crawling and third-party application for evidence collection	More data can be collected using currently available advanced technology
Hubert, K.	Evidence Collection from Social Media Sites	Social media provides evidence and how it is collected is critical	Evidence collection is tedious but necessary
Huey, L.	This is Not Your Mother's Terrorism: Social Media, Online Radicalization and the Practice of Political Jamming	Extremist groups use social media to conduct political jamming	Extremist groups use social media to recruit and influence members

(continued)

Table 1 (continued)

Author	Title	Key findings	Analysis
Jang, YJ., & Kwak, J.	Digital Forensics Investigation Methodology Applicable for Social Network Services	Proposed safe method of collecting evidence from social media	Proper evidence collection ensures legal admissibility
Karabiyik, U., Canbaz, M. A., Aksoy, A., Tuna, T., Akbas, E., Gonen, B., & Aygun, R. S.	A Survey of Social Network Forensics	There are many crimes that can occur over a social network	Before law enforcement can understand these crimes, they need to understand the network and how expansive it is
Kim, Y.	The Contribution of Social Network Sites to Exposure to Political Difference: The Relationships Among SNSs, Online Political Messaging, and Exposure to Cross-Cutting Perspectives	Social media influences individuals' exposure to political difference	Online political messaging affects social media users' political exposure
Kwahk, K., & Ge, X.	The Effects of Social Media on E-Commerce: A Perspective of Social Impact Theory	Social media interactions and social media commitment affect social influence	Social influence impacts e-commerce
Marsoof, A.	Online Social Networking and the Right to Privacy: The Conflicting Rights of Privacy and Expression	Sensitive information is often broadcasted on social media	There are limits to the laws in place that protect our privacy when it comes to social media
Murphy, J. P., & Fontecilla, A.	Social Media Evidence in Government Investigations and Criminal Proceedings: A Frontier of New Legal Issues	Social media creates inordinate data	Governments have tools to discover and obtain evidence
Parveen, F., Jaafar, N. I., & Ainin, S.	Social Media Usage and Organizational Performance: Reflections of Malaysian Social Media Managers	Organizations use social media for many reasons	Organizational use of social media impacts users

(continued)

Table 1 (continued)

Author	Title	Key findings	Analysis
Ramwell, S., Day, T., & Gibson, H.	Use Cases and Best Practices for LEAs	Best practices for undercover police operations and gathering open source information	Social media evidence can be collected inexpensively
Rubio, G.	Social Media Evidence, Digital Forensics, and Family Law	There are many laws that govern the use of social media	Policies must be followed for social media evidence to be acceptable
Seigfried-Spellar, K. C., & Leshnay, S. C.	The Intersection Between Social Media, Crime, and Digital Forensics: #Whodunit?	Social media plays a role in various crimes	Social media evidence can be extracted using digital forensics
Taylor, M., Haggerty, J., Gresty, D., Berry, T., & Almond, P.	Forensic Investigation of Social Networking Applications	When dealing with the misuse of social media, law enforcement may encounter jurisdictional boundaries	Procedures must be followed when dealing with digital evidence
Trottier, D.	A Research Agenda for Social Media Surveillance	Social media collects and disseminates personal information	Law enforcement should take advantage of this visibility
Uncel, M	"Facebook is Now Friends with the Court": Current Federal Rules and Social Media Evidence	Information found on social media can be viable evidence	In order to be used in court, social media evidence must overcome the challenges of discoverability, authenticity, hearsay, and reliability
Walnycky, D., Baggili, I., Marrington, A., Moore, J., & Breitinger, F.	Network and Device Forensic Analysis of Android Social-Messaging Applications	Data from social media is often stored unencrypted on smartphones	Data can be easily analyzed
Xiang, Z., & Gretzel, U.	Role of Social Media in Online Travel Information Search	Search engine results often list social media for travel-related searches	Social media has spread into all industries

3   Areas to Investigate Social Media Data

Social media can be used in many ways to aid digital forensics investigations. It can be used to link events together to create a timeline of the crime, to investigate an individual or group, or to demonstrate an underlying cause or circumstance. Social media may contain data that can be used to identify an individual or group and can illustrate the role played by this individual or group in an event. Such data may include user profiles, media uploads, user interests, and other content created either intentionally or unintentionally by the social media user. Social media may also contain underlying data such as GPS locations and timestamp information. Law enforcement can use such data when conducting criminal investigations. Other agencies such as the Internal Revenue Service and welfare agencies may also scour social media for data of interest. Social media can reveal a lot of personal details about a user. These details can be used to link the user to an event that is being investigated. Some avenues where social media can be used to the advantage of digital forensics investigations include using social media as a tool of surveillance, and as an evidence collection vector for criminal investigations and cases of terrorisms.

3.1   Undercover Police Operations

Social media can be used as a means of surveillance. Surveillance can be done in an on-going manner or from a retrospective view. By using digital forensics tools and techniques to review the information captured on social media, inferences can be made about a subject or an event. Law enforcement may utilize this use of social media. In particular, police undercover operations may find social media to be an essential tool. Oftentimes, profiles of interest will not have to directly interact with the undercover officer's profile; as this can be sometimes seen as unethical. Officers can simply browse the publicly viewable content offered by the profile being investigated. "Social media are now central to the visibility of personal information" [28]. The ease of access of this information creates various uses for it. For example, according to Fianyi [7], digital forensics tools can be used to detect and prevent crimes before they can be committed. This can be done by analyzing a user's content for clues of the premeditated crime. Clues may also be derived from engaging with the online user through the use of an undercover profile. However, as Ramwell et al. [24] explain, this must be done carefully as to not overstep into unlawful entrapment.

3.1.1   Data Removal

Users will interact more freely with a profile perceived as another normal user than a profile that is known to be operated by a police officer. A difficult part of creating undercover social media accounts is making the profiles believable. This means that the account must look and function just as any other profile. Undercover profiles will need to have a history to be perceived as real. The majority of social media profiles will contain personal details and histories as well as pictures and other distinguishing attributes. This exposes the officers to the threat of their personal information being gathered by malevolent individuals. Officers must also create protection for themselves by hiding their real personal accounts as this can blow their cover. Officers should engage in social media as little as possible to avoid this threat. However, this is often a difficult task as many officers would have already had and added content to personal social media profiles. Digital forensics can be conducted in a reverse manner to remediate this threat. Instead of capturing information from social media for analysis, the information can be sought out and removed. Cai et al. [4] suggested that public data from social media can be used to predict sensitive information about users and those involved with the profile. This data must be sanitized to remove the threat and safeguard other persons who could be connected to the officer.

3.2 Digital Footprints and Tagging

The avid use of social media amongst both younger and older generations today has created a multitude of online existence. Persons may be exposed to social media before they are even born, such as when a pregnant mother shares her ultrasound images. This example illustrates one of the countless ways that a person may create what can be called a "digital footprint". A digital footprint includes any trace of content online that can point to a specific individual. Although analyzing digital footprints can include common computer forensics, such as reviewing browser history and files saved on a computer, it can also include reviewing social media content. A recent example where social media was used to map the identities of users was the Facebook and Cambridge Analytica aspersion. It was reported that the political consulting firm, Cambridge Analytica, "gained access to private information on more than 50 million Facebook users" [9]. The firm harvested data from the social media site in an effort to use it to influence American voter behavior. This use of social media data has received tremendous backlash and is seen by some as an invasion of privacy. Big companies including social media companies engage in data mining. This practice allows the companies to learn and store detailed information about customers which is then sold to advertisers. These records of digital activity exhibit the customers' digital footprint and can unveil personal details and interests. These data can reveal hidden motives or other information that can be useful within a digital forensics investigation. Therefore, various paths of investigation can be created by surveilling an individual's digital footprint as created from social media sites and applications.

Individuals may obtain a digital footprint even without their own knowledge. "Tagging" features on social media sites and applications allow individuals to create traces of other people. "This is particularly due to the fact that social media encourage online consumers to be actively engaged in organizing the contents through activities like "digging" and "tagging", which, in turn, automatically create an ever growing link structure on the Internet" [31]. However, persons may also remove tags of themselves from posts made by others. This may be done in an attempt to hide evidence. Digital forensics investigations should include looking for such covert activity. If "untagging" activity is found, it may be concluded that an individual was trying to avoid such evidence from being used against him or her.

A tag can be used to name a person in a photo or message. A tag could also be location-based using the GPS functionalities within the social media application. This creates another investigative avenue. Location-based tags can be used to help locate persons of interest or a missing person. Tags used on social media can demonstrate a timeline of events. A photo tag can be used as evidence that an individual was within a certain area during a specific time. Digital forensics can be done on the uploaded photo to determine more information such as if the image was tampered with before it was uploaded. Doctored images can lead to falsified evidence and can ultimately determine the output of a case. Digital forensics must be conducted to ensure the originality of an image before it can be entered as evidence. Other than photo tags and location tags, there are also affiliation tags such as when a person is to be recognized as part of an association or club. This gives investigators another lead to exploit. In addition, even if a user has utilized privacy settings on his or her own account, the content may still be visible via another account that uses tags. This benefits an investigation by adding yet another sequence to follow. These connections to other individuals and groups can be examined to gather more information.

3.3   Open Source Intelligence

Using social media to gather information is sometimes viewed as an area called "open source intelligence". This utilization of social media is seen as open source because the information gathered is usually publicly available. This information can include anything that can be viewed on the user profiles of persons of interest. At first thought, this may not be expected to be a lot of information. On the contrary, many social media users are not aware of just how much information about them is publicly shared amongst platforms. "Social networks also encourage the publication of personal data, such as age, gender, habits, whereabouts, and schedules" [2]. Other personal information such as a date of birth, family relationships, and personal interests can be found on many online social media profiles. This information can be obtained to create a lead during an investigation. Open source intelligence is also highly beneficial to law enforcement because of its inexpensive cost and efficiency of collection. "Even without having to seek a warrant from the court or issue a subpoena, there are troves of social media evidence publicly available" [22]. This saves time and helps speed up the process of a digital forensics investigation since data can be collected by law enforcement for analysis fairly quickly.

Additionally, social media applications are now becoming integrated with each other and this poses another avenue for evidence collection. For example, popular social media platforms Facebook and Instagram allow user accounts to be linked. This connection effectively shares the information collected or posted on one application with the other application. What many users do not think about is the legal aspects of this connection. Facebook may have different policies for cooperating with law enforcement than Instagram does. However, because the content is shared across both applications, each platform could choose what to do with it. One company may share the same information that the other company is refusing to hand over. This gives law enforcement the upper advantage during digital forensics investigations. If useful evidence cannot be obtained from one platform, another one can be sought out. This is usually the case as most social media users will use more than one application to explore the different capabilities of each one. Therefore, the information is publicly shared and accessible, and can be collected through a variety of methods.

Although most social media sites and applications have security built in as well as additional privacy options for users, these are not commonly recognized. Many users are not aware of the enormous amount of information that can be gathered about them using social media. "The wealth of personal information uploaded to these websites makes it possible for cyber criminals to manipulate this information to their advantage and use it to commit criminal acts" [2]. However, Ramwell et al. [24] proclaim that this negligence works to the advantage of digital forensics investigations. Investigators will be able to collect this vast amount of publicly accessible evidence and be able to further examine content without legal or privacy restrictions. This aids an investigation because information can be gathered and processed with more efficiency. It also helps to supply the traces and timelines for a case. "Social networks open up a whole new world of information, because at least as much value is contained in the relationships between entities as in the entities themselves" [3]. Henceforward, using social media as a surveillance tool earns many benefits.

3.4 Criminal Investigations

Much as the use of social media amongst friendly users has expanded, so has the use amongst criminal organizations. Many criminal groups use social media as a medium for which to connect to and communicate with their members. Social media can allow the members of criminal groups to assume incognito identities. The members may use a different name on the social media profile, and they may not upload any identifying pictures. Social media allows the groups to create networks amongst their members as well as other criminal groups. The groups will use social media to collaborate activities. Digital forensics investigations will look into these connections established by the groups and will also consider the content they interact with. "Facebook has gained recent attention in the media for its use in aiding criminal investigations and college disciplinary hearing" [8]. Al-khateeb et al. [1] states that using seed knowledge such as known accounts and keywords used by the groups can help to discover these relationships. These groups use social media to coordinate attacks. However, analysis of these social networks can help to monitor their activities. Collecting data sets such as coordination strategies and interactions amongst members can help to track their behavior. The data created by criminal groups on social media must be considered when investigating criminal cases. "Furthermore, as social networks continue to replace traditional means of digital storage, sharing, and communication, collecting this type of data is also fundamental to the area of digital forensics" [14].

Social media content may also contain a large amount of metadata. If content from social media is captured as evidence, its metadata can be analyzed during a digital forensics investigation. The metadata found may include location, time, and author information. Al Mutawa et al. [2] found that using the Facebook application on iPhones and Android phones created such metadata. iPhones and Android phones currently dominate the smartphone market and are frequently investigated. The information retrieved from these devices using digital forensic tools can be entered as digital evidence. These data may also contain other information such as the audience of the post and its associated emotion. Social media applications usually allow a user to control the audience of the content he or she may post. The audience can typically be restricted to "only friends" or public. Restricted content may only be viewable by other users that the author interacts with. Public posts can usually be searched for and can be viewed by anyone who is also using the social media application. If content is restricted, the user will assume only the allowed viewers can see the content and so the digital forensics investigator must also assume this. However, the digital forensics investigator can sometimes also find out what other user accounts viewed the data. If content is "liked" by another user, it is perceptibly assumed that this user viewed the data. If that content is related to the criminal act, this user will have a hard time denying knowledge of the act. Therefore, investigators may be able to analyze social media content metadata to provide expert observations during a criminal case.

3.4.1 White Collar Crimes and E-Commerce

White collar crimes are those that may not include violence or physical threats but are financially motivated and usually conducted by businesses and governments. White collar crime also includes fraud and money laundering. The explosion of social media has provided new courses for these crimes. Many businesses use social media to promote their significance and communicate with the public. Numerous companies now conduct business online to adhere to the social trends being developed. Research by Kwahk and Ge [20] suggests that social media interactions affect social influence which in turn impacts purchase intentions in e-commerce. Social influence can persuade others to make similar purchases or visit similar e-commerce websites. Online shopping has become a permanent fad and customers enjoy its ease and efficiency. Businesses offer these services to keep up a competitive advantage. Parveen et al. [23] state "that social media can provide a competitive advantage when the organizations use it innovatively and differently from their competitors". With this increasing pressure from customers, businesses may overlook the security that is essential to conducting business online. Thus, many businesses experience attacks that compromise their systems. Conducting business online, or e-commerce, poses many risks to a company. "Some of the risk that are associated with ecommerce includes but not limited to, stolen credit card number, personality theft, stolen Social Security Numbers, and sometimes goods bought and paid for never arrives" [7]. The massive number of criminal acts that can be completed using social networks exemplifies the significance of understanding digital forensics in this area.

Criminal cases frequently involve a financial motive. Hackers may try to steal data in order to sell it for profit. Social media continues to expand its capabilities and functions with almost every constant update. These applications often have advertisements and in-app purchases. Users are usually made aware of this before downloading the application. However, the incentive to allow these advertisements and in-app purchases is that users may receive additional features for use within the application. These in-app purchases are a form of e-commerce that is sometimes called microtransactions. They are called microtransactions because the purchases are usually small in price. Nonetheless, the purchase involves the use of some kind of payment mechanism. The payment choice for a majority of users will be a credit card. Parveen et al. [23] suggest that credit card campaigns on social media platforms such as Facebook can generate revenue for the company. Credit card data is frequently a target of hackers with a financial motive. Credit card data can be sold to other online criminals, or sometimes replicated into a physical card and then sold. Hackers have new knowledge and cleverer ways to fulfill their financial motives. Credit card data thieves will act fast. Usually the credit card owner will not be alerted until fraudulent purchases have already been made. Banks have come up with new ways to try to detect fraudulent purchases, but criminals are equally finding new ways of exploitation. Thieves are not only targeting individual credit card owners, but also large enterprises. News stories repeatedly report on data breaches in large organizations where thousands of customer credit card data is stolen. Large companies will usually invest the time and money to conduct a comprehensive digital forensics investigation to understand how the breach occurred, how much data was compromised, and how to prevent future occurrences. The digital forensics investigator must therefore be able to analyze the social media application to provide sound conclusions. The advantage here is that with this gained knowledge, security can be further developed and social media can become more protected.

3.4.2 Online Chats

Online social media games are now a way that users can interact as well. Many online social media games include a chat feature. The chat feature allows users to engage in conversation and conduct other interactions. Kim [19] suggests that online political messages such as those within online chats can result in inadvertent exposure to political differences. This can influence users. Criminals may also use this online chat feature because the chats are usually closed out once the user stops using the application. They think that the data and its traces have disappeared. However, some applications may actually save this data. Jang and Kwak [17] assert that social media conversations between users can be retrieved from a computer's volatile memory. The memory will store the content of the conversation if a web browser was used to access the social media site. As another digital trace, this conversation can be retrieved using digital forensic tools. The conversations may also be saved temporarily on the servers of the social media provider. The length of time that the chat is saved for depends on the social media company. The digital forensics investigator may have to act swiftly to recover these conversations. An analysis can be done once the conversation is recovered from the servers of the social media company or from the suspect's machine. The analysis can investigate recipients, content, and times of the conversation. "Electronic evidence retrieved from social networking activities on a suspect's machine can be of great assistance in investigating a criminal case by incriminating or proving the innocence of a suspect" [2]. This data can then be presented in the court of law as evidence. Sometimes, the social media providers can provide helpful tools to discover and extract such data. Using tools supported by the provider can help to ensure its integrity. The developers of these tools can also be entered as expert witnesses, helping to solidify a criminal proceeding.

3.5   Online Terrorism

Terrorism can be defined as using violence or the threat of violence to coerce and intimidate a nation or its people to achieve political or ideological objectives. Online terrorism, or cyberterrorism, is using the Internet as a vehicle to achieve these objectives. Acts of online terrorism can include widespread denial of service attacks such as those that disable an entire power grid. Online terrorism can also include introducing viruses and other malware into computer systems or networks, typically government-backed networks.

Online terrorism can undoubtedly be conducted using social media. Since social media is so widespread, it offers a path of mass communication. A study by Huey [16] states that the practice of political jamming is used to popularize pro-terrorist messages. Social media users become attracted to these messages by the perception that they are funny or facetious. This communication can then give extremist groups more attention and influence. Terrorist groups may use social media to display their beliefs and other messages and post content to exclaim their ideologies or activities. Social media must then be filtered. Many applications will have policies and procedures in place for when its providers are allowed to remove a user's content or deactivate the user's account. This is solely at the discretion of the social media company. However, digital forensics can be utilized when content is not removed and is recoverable. The content and postings of these terrorist organizations can be analyzed along with metadata and other information found on the application or site. This delivers another stipulation in the case of online terrorist attacks involving viruses and other malware. In the United States of America, there are laws that prohibit the unauthorized access of computers and computer systems. One such law is the Computer Fraud and Abuse

Act. Governments may track down a suspected hacker or cyberterrorist while working under the coverage of national security. Digital forensics will be used to conduct such traces, and social media activity will provide valuable information about the accused. An investigation will yield information about the terrorist group that can be further analyzed and presented in a court of law. The evidence gathered on social media may be used to understand how the attack occurred and who else was involved. Murphy and Fontecilla [22] discuss a criminal case where federal authorities relied on social media. The case involved four persons believed to be associated with the Al Qaeda. The investigators utilized an undercover online presence to "elicit damaging statements from the defendants, recorded Skype conversations between a confidential informant and the defendants, and relied on the social media content that each defendant "liked," "shared," or on which the defendant commented" [22]. These undercover operations take the advantage of social media to help law enforcement capture terrorists and may lead to disabling the group.

3.5.1 Recruitment

Social media presents a new front for terrorist organizations to conduct recruitment. These organizations may use social media to coerce individuals into joining the organization. A terrorist organization may research a potential recruit using social media to find out personal details. It will then use these details to inflict fear into the recruit and force him or her into conducting an act that is beneficial to the organization. Nonetheless, the situation may also be reversed. Individuals may use social media to initiate contact with terrorist organizations. They will use social media as a tool of communication whereby they can interact with the organization. Huey [16] says "one study of U.S. Al-Qaeda recruits found that these individuals were typically confused, young people searching to define themselves and gain a sense of purpose". These impressionable youths are often targeted, and this naivety will be taken advantage of by the extremist group to instill its beliefs. Al-khateeb et al. [1] discuss how one such terrorist organization, the Islamic State of Iraq and the Levant (ISIL), use Internet forums and social media to influence vulnerable youth to join the organization and participate in its acts. This example shows that social media can facilitate unlawful acts, but digital forensics can be used to analyze these patterns and potentially eliminate the issue.

4 Adversities Against the Use of Social Media as Evidence

As mentioned throughout this paper, social media can be a very powerful asset in any digital forensics investigation. It can provide a way to link what may seem like unrelated evidence, and in many cases is the key component used to indict a suspected criminal. As powerful a tool as it may be, it is not without its drawbacks. How that evidence is collected and preserved is very important and law enforcement may encounter various legal barriers during these processes. If appropriate procedures are not followed at any given point in an investigation, this discrepancy may compromise or invalidate the social media evidence. Thus, the evidence may become inadmissible in a court of law. In addition to these legal issues, there are also numerous ethical implications of using social media data as digital evidence that need to be considered.

4.1 Legal Barriers

How social media evidence is used in a court of law, both at the federal-level and the state-level, is heavily regulated. Since the use of social media and other forms of electronic communication are constantly evolving, there must be rules in place to regulate them. Such existing laws are not clear cut, and legislators must often use precedence to determine social media's role in digital forensics investigations or to justify how law enforcement can apply these rules in an investigation. Laws and statutes in place today that affect this use of evidence include the Electronic Communications Privacy Act which consists of the Title I Wiretap Act and the Title II Stored Communications Act. The Federal Rules of Civil Procedure is another current legislation that limits how law enforcement uses digital evidence; including social media evidence.

4.1.1 Title I of the Electronic Communications Privacy Act

Passed in 1986, the Electronic Communications Privacy Act was enacted in an attempt to update the current statute that was in place pertaining to Title III of the Omnibus Crime Control and Safe Streets Act of 1968, otherwise known as the "Wiretap Statute" [25]. The Act of 1968 applied to hard telephone lines, but with the new update of 1986 it became inclusive of all electronic communications. The Title I Wiretap Act is in accordance to 18 U.S.C. § 2511–2522 which disallows the interception of "any wire, oral, or electronic communication" [25].

According to Rubio [25], Section 2515 states:

> Whenever any wire or oral communication has been intercepted, no part of the contents of such communication and no evidence derived therefrom may be received in evidence in any trial, hearing, or other proceeding in or before any court, grand jury, department, officer, agency, regulatory body… or other authority of the United States, State, or a political subdivision thereof if the disclosure of that information would be in violation of this chapter.

There is an exception to this rule. One instance where this evidence could be submitted is when it is proven that all other methods to obtain the evidence have been exhausted and that the highest level of legal scrutiny was enacted. In addition to these stipulations, a warrant or court order must also be presented.

Chapter 18 of the United States Constitution § 2511 [2] (A) states:

It shall not be unlawful for a provider of wire or electronic communication service… to intercept, disclose, or use that communication… which is a necessary incident to the rendition of this service or to the protection of the rights or property of the provider of that service.

Therefore, it is in the service providers' (i.e. Facebook, Twitter, Instagram) best interests to cooperate with law enforcement and disclose its user's information or to provide an avenue for law enforcement to conduct an investigation. In this case, the provider will not be penalized.

4.1.2 Title II of the Electronic Communications Privacy Act

Although Title II, otherwise known as the Stored Communications Act, provides protection for service providers, it also outlines various rules and laws that must be obeyed by these social network providers. It should also be noted that these laws and statutes do not only apply to social media network services but also to Internet Service Providers and citizens of the public.

Rubio [25] provides that the laws outlined in Section 2702 are:

1. a person or entity providing an electronic communication service to the public shall not knowingly divulge to any person or entity the contents of a communication while in electronic storage by that service; and
2. a person or entity providing remote computing service to the public shall not knowingly divulge to any person or entity the contents of any communication which is carried or maintained on that service.
   a. on behalf of, and received by means of electronic transmission from… a subscriber or customer.
   b. solely for the purpose of providing storage or computer processing services… if the provider is not authorized to access the contents of any such communications for providing services other than storage…
3. … shall not knowingly divulge a record or other information pertaining to a subscriber to or customer of such service… to any governmental entity.

In accordance to Chapter 121 of 18 U.S.C § 2701–2712, Title II of the Electronic Communications Privacy Act disallows the unauthorized access of "a facility through which an electronic communication service is provided" [25]. This means that unless law enforcement has clearance, the authority of a search warrant, or a court order, any potential evidence from the social media provider does not need to be handed over. However, if an investigator can secure a warrant, then outside parties can be compelled to cooperate.

4.1.3 Federal Rules of Civil Procedure

If it is discovered that social media evidence can be used in the court of law after it is handed over to the authorities, there are various rules and procedures that must be followed. Rubio [25] states that the Federal Rules of Civil Procedure (FRCP) regulates how such cases are presented before civil courts. While these rules address electronically stored information, they do not explicitly mention the use of social media evidence. However, in 2006 the courts ruled that such evidence must meet the same requirements as traditional evidence such as being carefully handled and documented. Among those requirements, social media evidence must be relevant, reliable, sufficient, and cover the burden of proof required for the case.

If established procedures are not followed at any given point in an investigation, it can cause the evidence to become compromised or invalidated. In such a case, any social media evidence or digital evidence becomes inadmissible in court. Rule 37 (e) of the FRCP states that electronically stored information should be preserved in the anticipation or current proceeding of a lawsuit. The FRCP also explains that if any data is lost because steps were not taken to preserve the evidence, that same data cannot be replaced. If it is discovered that prejudice was intently established in court by either party, then the trier of facts must assume the compromised information was unfavorable, dismiss the action, or enter a default judgement. This is why preservation of evidence is so important in digital forensics, and why social media evidence can be a vulnerable asset in an investigation. As a result, whenever digital evidence is involved in a crime or investigation, digital forensics investigators must step in and carefully follow a standardized investigative plan. This plan can become very tedious when such evidence is to be collected. However, investigators must preserve the integrity of the evidence collected in order for it to be admissible. This is done by establishing a clear chain of custody, proper documentation, careful gathering, control, storage, and processing of the evidence. Each step of the plan must be carried out cautiously and meticulously.

4.1.4   A Global Crime Scene

The laws and regulations described previously apply to cases and investigations conducted within the United States of America. What happens when the crime scene becomes global? Do the same rules apply to users of a social networks that are so far reaching? Can law enforcement still use the evidence if it is stored in other countries? This depends on the jurisdiction, and law enforcement must undertake a case by case basis. However, it can be assumed that other countries have policies and legislations in place to deal with crimes involving social media evidence. One example of foreign legislation is the Australian High-Wire Act: Cyber-Safety and the Young. This policy was put in place as an attempt to protect the youth that are online from terrorist recruitment and other immoral activities such as sex-trafficking and abuse. This protection is done by monitoring and flagging the Internet activity of users [6]. Another foreign policy in place is the UK Regulation of Investigatory Powers Act. This law allows the monitoring or collection of data in the case that there is misuse of social networking applications within an organization. Much like U.S. investigatory laws, the UK Regulation of Investigatory Powers Act calls for a court order or search warrant before law enforcement can become involved. Therefore, monitoring and prevention go hand in hand.

4.2   Ethical Implications

What society dictates as ethical and unethical varies from individual to individual or in legal proceedings, case by case. What might be acceptable to one person may not be seen that way in the eyes of another. For example, why is it considered okay for a police officer to lie and create a fake social networking profile, but frowned upon when an ordinary individual does this? Most people are taught from a young age that telling the truth is always better than lying. However, some individuals may think that lying is only a serious problem when it is assumed that criminal activity is occurring. In the case of undercover police operations, why is it suddenly okay? These are the questions that digital forensics investigators, lawmakers, and critically thinking individuals must ask themselves. The anonymity that social media platforms provide enable any user to act in any way and assume any identity. This applies not only to general users, but law enforcement as well. Where is the line drawn? These methods can be justified when a case can be successfully closed without putting any innocent individual's life at risk.

4.2.1   The Fourth Amendment

Another question raised in dealing with social media evidence is whether or not an individual who uses the platform owns what is posted or communicated online. Can law enforcement freely use what a suspect posts without penalty? What about the right to privacy? The Fourth Amendment of the United States Constitution is what gives this right to privacy and freedom from unlawful search and seizure.

The Fourth Amendment explicitly states:

> The right of the people to be secure in their persons, houses, papers, and effects, against unreasonable searches and seizures, shall not be violated, and no warrants shall issue, but upon probable cause, supported by oath or affirmation, and particularly describing the place to be searched, and the persons things to be seized.

Most, if not all, social media networking sites have a "Terms and Conditions" policy that they require users to consent to before using their services. It has been determined that data shared on social media with other users is not allowed the privilege of being protected by the Fourth Amendment because that data is made public as soon as it is distributed or posted. This visibility is an advantage for digital forensics investigations, but not so much one for the individual or individuals on the other side that are being investigated.

5   Conclusion

Social media evidence has the potential to be a powerful asset in any investigation using digital forensics, so the advantages far outweigh the disadvantages for now.

The evidence can be obtained using a variety of methods including open source intelligence, analyzing metadata, undercover police operations, and many more. Since many crimes involve social media, the need for legalities and ethics when handling this evidence is very critical. Technology is innovative and how individuals adapt to that innovation will play a big role in the generations to come. Social media is constantly evolving and with it, how individuals perceive the new technological society. The laws in place now must evolve as well to adjust to these changes. Where ethics and morals are questioned, society must face the reality that issues considered right or wrong will also change.

References

1. Al-khateeb, S., Conlan, K.J., Agarwal, N., Baggili, I., Breitinger, F.: Exploring deviant hacker networks (DNH) on social media platforms. J. Digit. Forensics Secur. Law **11**(2), 7–20. https://login.libweb.lib.utsa.edu/login?url=https://search-proquest-com.libweb.lib.utsa.edu/docview/1825888188?accountid=7122
2. Al Mutawa, N., Baggili, I., Marrington, A.: Forensic analysis of social networking applications on mobile devices. Digit. Investig. **9**, S24–S33 (2012). https://doi.org/10.1016/j.diin.2012.05.007
3. Bradbury, D.: In plain view: open source intelligence. Comput. Fraud Secur. **2011**, 5–9 (2011). https://doi.org/10.1016/S1361-3723(11)70039-2
4. Cai, Z., He, Z., Guan, X., Li, Y.: Collective data-sanitization for preventing sensitive information inference attacks in social networks. IEEE Trans. Dependable Secure Comput. https://doi.org/10.1109/tdsc.2016.2613521, 26 Sept 2016
5. Carrier, B. (2005, March 17). File System Forensic Analysis. Pearson Education Inc., Upper Saddle River, NJ
6. Dean, G., Bell, P., Newman, J.: The dark side of social media: review of online terrorism. Pak. J. Criminol. **3**(3), 107–126. https://www.researchgate.net/profile/Frederic_Lemieux/publication/236730770_Assessing_Terrorist_Risks_Developing_an_Algorithm-Based_Model_for_Law_Enforcement/links/02e7e535bef3516b60000000/Assessing-Terrorist-Risks-Developing-an-Algorithm-Based-Model-for-Law-Enforcement.pdf#page=117
7. Fianyi, I.D.: Curbing cyber-crime and enhancing e-commerce security with digital forensics. Int. J. Comput. Sci. Issues (IJCSI) **12**(6), 78–85. https://search-proquest-com.libweb.lib.utsa.edu/docview/1752642603?pq-origsite=summon
8. Gosling, S., Gaddis, S., Vazire, S.: Personality impressions based on Facebook profiles (n.d.). https://pdfs.semanticscholar.org/47b3/db84c94dac4098163f0fb6886a8ffea0a2fc.pdf
9. Granville, K.: Facebook and Cambridge analytica: what you need to know as fallout widens, 19 Mar 2018. https://www.nytimes.com/2018/03/19/technology/facebook-cambridge-analytica-explained.html
10. Grimm, P., Bergstrom, L., O'Toole-Loureiro, M.: Authentication of social media evidence. Am. J. Trial Advocacy, **36**. http://www.atyvideo.com/documents/American%20Journal%20of%20Trial%20Advocacy%20Authentication%20of%20Social%20Media%20Evidence.pdf
11. Hesketh, I., Williams, E.: A new canteen culture: the potential to use social media as evidence in policing. Policing: J. Policy Pract. **11**, 346–355. https://doi.org/10.1093/police/pax025
12. Hoffmeister, T.: The challenges of preventing and prosecuting social media crimes. Pace Law Rev. **35** (2014). https://digitalcommons.pace.edu/cgi/viewcontent.cgi?referer=https://www.google.com/&httpsredir=1&article=1877&context=plr
13. Holt, M., San Pedro, V.: Social media evidence: what you can't use won't help you. Fla. Bar J. (2014). https://www.floridabar.org/the-florida-bar-journal/social-media-evidence-what-you-cantuse-wont-help-you-practical-considerations-for-using-evidence-gathered-on-the-internet/

14. Huber, M., Mulazzani, M., Leithner, M., Schrittwieser, S., Wondracek, G., Weippl, E.: Social snapshots: digital forensics for online social networks (n.d.). https://www.sba-research.org/wp-content/uploads/publications/social_snapshots_preprint.pdf

15. Hubert, K.: Evidence collection from social media sites (2014). https://www.sans.org/reading-room/whitepapers/legal/evidence-collection-social-media-sites-35647

16. Huey, L.: This is not your mother's terrorism: social media, online radicalization and the practice of political jamming. J. Terror. Res. **6**, 25 May 2015. http://doi.org/10.15664/jtr.1159

17. Jang, YJ., Kwak, J.: Digital forensics investigation methodology applicable for social network services. Multimed. Tools Appl. **7**(14), 5029–5040 (2015). https://doi-org.libweb.lib.utsa.edu/10.1007/s11042-014-2061-8

18. Karabiyik, U., Canbaz, M.A., Aksoy, A., Tuna, T., Akbas, E., Gonen, B., Aygun, R.S.: A survey of social network forensics. J. Digit. Forensics Secur. Law, **11**(4), 31 Dec 2016. https://commons.erau.edu/cgi/viewcontent.cgi?article=1430&context=jdfsl

19. Kim, Y.: The contribution of social network sites to exposure to political difference: The relationships among SNSs, online political messaging, and exposure to cross-cutting perspectives. Comput. Hum. Behav. **27**, 971–977 (2011). https://doi.org/10.1016/j.chb.2010.12.001

20. Kwahk, K., Ge, X.: The effects of social media on e-commerce: a perspective of social impact theory. Syst. Sci. (HICSS) (9 Feb 2012). https://doi.org/10.1109/hicss.2012.564

21. Marsoof, A.: Online social networking and the right to privacy: The conflicting rights of privacy and expression. Int. J. Law Inf. Technol. **19**, 110–132 (2011). https://doi.org/10.1093/ijlit/eaq018

22. Murphy, J.P., Fontecilla, A.: Social media evidence in government investigations and criminal proceedings: a frontier of new legal issues. Richmond J. Law Technol. **19**(3) (2013). https://scholarship.richmond.edu/cgi/viewcontent.cgi?referer=https://scholar.google.com/&httpsredir=1&article=1380&context=jolt

23. Parveen, F., Jaafar, N.I., Ainin, S.: Social media usage and organizational performance: reflections of Malaysian social media managers. Telemat. Inform. **32**, 67–78 (2015). https://doi.org/10.1016/j.tele.2014.03.001

24. Ramwell, S., Day, T., Gibson, H.: Use cases and best practices for LEAs. In: Akhgar, B., Bayerl, P., Sampson, F. (eds) Open Source Intelligence Investigation, pp. 197–211. Springer, Cham (2016). https://doi-org.libweb.lib.utsa.edu/10.1007/978-3-319-47671-1_13

25. Rubio, G.: Social media evidence, digital forensics, and family law. https://www.protegga.com/wp-content/uploads/2016/01/Social-Media-Evidence-Digital-Forensics-and-Family-Law.pdf

26. Seigfried-Spellar, K.C., Leshnay, S.C.: The intersection between social media, crime, and digital forensics: #WhoDunIt?. In: Digital Forensics: threatscape and Best Practices, pp. 59–67. (2016). https://doi.org/10.1016/B978-0-12-804526-8.00004-6

27. Taylor, M., Haggerty, J., Gresty, D., Berry, T., Almond, P. Forensic investigation of social networking applications. Netw. Secur. **11**, 9–16 (2014). https://doi.org/10.1016/S1353-4858(14)70112-6

28. Trottier, D.: A research agenda for social media surveillance. Fast Capital. **8**(1) (2011). https://www.uta.edu/huma/agger/fastcapitalism/8_1/trottier8_1.html

29. Uncel, M.: "Facebook is now friends with the court": current federal rules and social media evidence. Jurimetrics **52**, 43–69 (2011). http://www.jstor.org/stable/23239803

30. Walnycky, D., Baggili, I., Marrington, A., Moore, J., Breitinger, F.: Network and device forensic analysis of android social-messaging applications. Digit. Investig. **14**, S77–S84 (2015). https://doi.org/10.1016/j.diin.2015.05.009

31. Xiang, Z., Gretzel, U.: Role of social media in online travel information search. Tour. Manag. **31**, 179–188 (2010). https://doi.org/10.1016/j.tourman.2009.02.016

The Way Forward

Xiaolu Zhang and Kim-Kwang Raymond Choo

Abstract This chapter presents a summary of this book. We also outline a number of future research agenda based on the failure and success gained during the two years of implementing experiential learning in both undergraduate level and graduate level digital forensic courses, with the hope that the findings reported in this book can be used for facilitating the practice-oriented digital forensic education in the future.

At the end of the semester, students were encouraged to provide feedback on their learning experience and how we can improve on our future delivery. Appended below are excerpts (in verbatim) from the students' evaluation for both the course and the instructor.

- *"Over the course of the semester, I did gain a lot of insight into the field of Computer forensics and it provided a lot of foundational [sic] knowledge on the subject as well as hands on experience with necessary tools. However, the class is quite challenging for beginners and would prefer if hands-on examples were performed in class and the tools available in the lab should be updated."*
- *"The most challenging aspect of this course were the lab assignments which required access to UTSA's CSL [cyber security lab] which operates on a very limited schedule: Mon.—Thurs. closing at 10PM, Fri. closing at 7PM, Sat. open from 10am–4pm, closed Sun. I inquired, but was told there is no remote access to CSL servers. This situation is especially difficult for graduate students who work during the day and take other evening classes (some of which also issue assignments requiring CSL access). The only extended window to work on assignments were the six hour slot on Sat. or evenings when class lecture was cancelled."*
- *"Super interesting and fun course. The concepts were presented well and real world scenarios were applied. The project should stay as part of the class, however it could be better defined in each milestone, etc. It's still very hard for undergraduates*

X. Zhang · K.-K. Choo (✉)
Department of Information System and Cyber Security,
University of Texas at San Antonio, San Antonio, TX, USA
e-mail: raymond.choo@fulbrightmail.org

X. Zhang
e-mail: xiaolu.zhang@utsa.edu

© Springer Nature Switzerland AG 2020
X. Zhang and K.-K. R. Choo (eds.), *Digital Forensic Education*,
Studies in Big Data 61, https://doi.org/10.1007/978-3-030-23547-5_15

to be able to perform research at the speed of the class however, so maybe scope could be reduced just a tiny bit."

- *"The course is incredibly technical and hands on which makes it very rewarding."*
- *"Instructor was very prepared and had clear expectations from the class. Very easy to work with and gave plenty of freedom for research and developing the research project. The research project relied heavily on one's knowledge of VM manipulation and analysis."*
- *"This course is a good course for the MSIT Cyber Security program. The hands-on exercises are very helpful, that's where I've learned the most."*
- *"This is a challenging course that is a cornerstone of the program. I am very glad that I took it."*
- *"I really enjoyed this class. It really challenged me to think in ways I haven't needed to before. I've always had an interest in this field and this has really helped."*
- *"Class is great too, lots of opportunities to learn different tools and methods for digital forensics. Also because of the project we got to simulate real world experience. This is definitely one of the most challenging courses I took at UTSA but I learned a lot in it and it applies to courses outside of forensics as well."*
- *"It was a really great experience and I got a ton out of it. Learning to read hex and file systems is something that's been on my to do list, and I hope to continue to learn through my job and further learning. Also thanks again for the opportunity to try and publish a paper with you, we all appreciate it a lot."*

We will also outline some of the challenges we faced during the courses, and the associated research opportunities.

- Access to the commercial forensic tools is limited to the opening hours of the laboratory, as pointed out by several students. While students were encouraged to explore the use of open source forensic tools, the latter generally has limited functionalities and features.

  – This also highlights the need for more research and development efforts into designing open source forensic tools, with comparable functionalities and features to commercial forensic tools, for students and researchers in institutes of higher education.

- The amount of time required by groups working on research assignments to generate the datasets required for analysis. For example, for students working on dating app analysis, they would first need to install the various dating apps on their mobile devices, set up a number of profiles for research, interact among the research profiles (students were advised not to reply to messages sent by other dating app users and/or messaged other dating app users), and conduct other activities typical of a dating app user but without interacting with other dating app users.

  – This highlights the need for making available datasets (such as those generated by the groups described in Part II of this book) that can be shared securely among students and researchers in participating institutes of higher education, using approaches such as blockchain [1, 2, 4].

- Lack of an information and knowledge-sharing platform that allows groups to learn from prior experience, including from student groups in past semester(s)/year(s) working on similar devices and/or apps. For example, as explained in our earlier work, *"in a typical investigation process, two or more investigators located in different cities and/or countries may be forensically examining the same (type of) device at the same time. As the experience and background of both investigators are likely to vary, the outcomes of the forensic investigations may also differ (e.g. in terms of the types and extent of artifacts being recovered). [.... However,] due to the sensitive nature of data acquired from forensic examination, data is seldom shared."*.]

  - This highlights the need for having a *"forensic knowledge-sharing platform designed to facilitate the sharing of knowledge/experience via a schema generated from another investigation of a similar device. Such a schema captures the knowledge of an investigator who had previously examined a particular device, and contains information, such as type and nature of artifacts that could be acquired, to guide another investigator's forensic examination"* [3].
  - In addition to the forensic knowledge-sharing platform, an open-source sharing platform, such as Syracuse University's "Hands-on Labs for Security Education",[1] where other digital forensic instructors can go to and download the lecture materials, hands-on lab exercises, datasets that students can use for analysis, etc, will benefit the community. In addition, such a platform can also be designed to allow other digital forensic instructors to share their experience and materials (e.g., datasets) with the community.

References

1. Banerjee, M., Lee, J., Choo, K.R.: A blockchain future for internet of things security: a position paper. Digit. Commun. Netw. **4**, 149–160 (2018)
2. Zhang, L., Luo, M., Li, J., Au, M.H., Choo, K.R., Chen, T., Tian, S.: Blockchain based secure data sharing scheme for internet of vehicles. Veh. Commun. **16**, 85–93 (2019)
3. Zhang, X., Choo, K.R., Beebe, N.: How do i share my iot forensic experience with the broader community? An automated knowledge sharing iot forensic platform. IEEE Internet Things J. (2019). https://doi.org/10.1109/JIOT.2019.2912118
4. Zhu, L., Wu, Y., Gai, K., Choo, K.R.: Controllable and trustworthy blockchain-based cloud data management. Future Gener. Comput. Syst. **91**, 527–535 (2019). https://doi.org/10.1016/j.future.2018.09.019

---

[1] http://www.cis.syr.edu/wedu/seed/index.html.

Printed in the United States
By Bookmasters